Roofing
the Right Way
Third Edition

Roofing
the Right Way
Third Edition

Steven Bolt

McGraw-Hill
New York San Francisco Washington, D.C. Auckland Bogotá
Caracas Lisbon London Madrid Mexico City Milan
Montreal New Delhi San Juan Singapore
Sydney Tokyo Toronto

Library of Congress Cataloging-in-Publication Data

Bolt, Steven, 1949-
    Roofing the right way / Steven Bolt. —3rd ed.
      p.   cm.
    Includes bibliographical references and index.
    ISBN 0-07-006649-3 (hc). — ISBN 0-07-006650-7 (pb)
    1. Roofing.   I. Title.
TH2431.B58   1997   96-41056
695—DC20   CIP

McGraw-Hill

A Division of The McGraw-Hill Companies

1 2 3 4 5 6 7 8 9 0 DOC/DOC 9 0 1 0 9 8 7 6

ISBN   hc   0-07-006649-3
           pb   0-07-006650-7

The sponsoring editor for this book was Zoe Foundotos, the editing supervisor was Scott Amerman, and the production supervisor was Don Schmidt. It was set in Century Schoolbook by Editorial Compuvision, Mexico.

Printed and bound by R. R. Donnelley & Sons Company

McGraw-Hill books are available at special quantity discounts to use as premiums and sales promotions, or for use in corporate training programs. For more information, please write to the Director of Special Sales, McGraw-Hill, 11 West 19th Street, New York, NY 10011. Or contact your local bookstore.

For Brian and Chad

Contents

Acknowledgments

This book is a product of my roofing experience and my interest in words. While it might appear that roofing and writing are dissimilar skills, I am fortunate to have colleagues and friends who have shown me that both disciplines require precision and craftsmanship.

I want to thank the following for their encouragement and support: Cindy Bolt, Ken Bolt, Barbara Brown, Dixon Brown, Daniel Clem, Tom Conner, Virginia Conner, Chet Dawson, Julia Dawson, Julie Dawson, Geraldine Eustice, Anna Farnkoff, Tom Farnkoff, Karen Schulze-Hall, Tim Hall, Douglas Hartz—who is with me on every page—Maurine Kent, Roy Kent, Ed McCarthy, Kip Naughton, Roland Phelps, Doug Robson, Carl Schulze, Helena Schulze, Christopher Surrey, Samantha Surrey, and Scott Surrey.

Thanks also go to Carolyn Anderson, Jackie Boone, Suzanne Cheatle, Jean Fiori, Susan Hansford, Patsy Harne, Richard Hawkins, Rita Henderson, Linda King, Lisa Meliott, Nadine McFarland, April Nolan, Charles Sanders, Kimberly Shockey, Joanne Slike, Kimberly Tabor, Leslie Wenger, Joan Wieland, and Bob Ziegler for lending their talent and patience.

Special recognition is due Ray Collins for providing the title alliteration.

I wish to thank Onduline, Inc., and Patricia Pitts for supplying the illustrations that appear in chapter 11, and I especially thank Greg Thorne for all of the other drawings that appear throughout the book.

Employees of Sporik Roofing of Steelton, Pennsylvania, Sunvek Roofing of Phoenix, Arizona, R.F. Fager Company, Mechanicsburg, Pennsylvania, and McGraw-Hill have been especially generous with their time and efforts.

The first edition of this book is dedicated to Deborah Shaw, with special thanks to John J. Romano. *Amícus certus in ré incertá cenitur.* The second edition of this book is dedicated to Shirley and Julie, and those with memories of them.

About This Book

Roofing the Right Way, Third Edition, describes the materials, tools, and step-by-step application techniques for the successful installation of roofing materials on a new building or the reroofing of an existing one. Comparisons and descriptions of the most popular types of roofing materials are included, as well as guidelines on how to determine when your asphalt roof needs replacement, do-it-yourself application details, and checklists for choosing a reliable contractor.

The book provides details on the advantages of using specialized roofer's tools and equipment and examines the most common roofing materials—asphalt fiberglass-based shingles, wood shingles and shakes, metal roofing, cement-tile roofing, and roll roofing—and how to waterproof roofline intrusions and obstacles such as chimneys, walls, valleys, and vents.

According to the Home Ventilation Institute, about 70 percent of all homes do not have adequate ventilation. Consequently, I've included installation material on roof ventilation devices such as turbines, ridge ventilators, and skylights.

If installing a new roof or reroofing over worn shingles seems an intimidating task, consider that you can save as much as two-thirds the cost of keeping your home waterproof by doing the shingling yourself. Glance through the chapters that describe the style or the application of the roofing materials you want to use on your home. You'll find that the work is often well within your capabilities. Step-by-step photographs and drawings guide you through the procedures.

Practical tips that provide advice based on years of professional experience are included. Descriptions and details for all of the roofing procedures are subdivided into easy-to-grasp segments so that you can readily discuss your roofing needs with contractors and suppliers.

An appendix provides concise answers to homeowners' most frequently asked roofing questions. Technical terms mentioned throughout the book are compiled in the Glossary. Addresses of equipment manufacturers and product suppliers are listed in Resources at the back of the book.

If you're uncertain about taking on your own roofing project, *Roofing the Right Way* can help you understand what it will take to get the job done. If you are certain you want to hire a contractor to roof your home, chapter 3 provides guidelines and checklists for how to choose a contractor and the details to discuss and list in a contract.

From the examples explaining how to estimate roofing materials to the definitions of technical terms, the emphasis is on carefully written and technically precise descriptions. With the aid of this book, you'll be informed and ready to discuss your roofing needs with material suppliers and prospective contractors.

Chapter **1**

Roof Conditions and Materials

According to United States Census Bureau statistics, about 23 million single-family homes are at least 50 years old, and several more million homes are at least 15 years old. With approximately 80 percent of American homes roofed with asphalt-based shingles or asphalt fiberglass-based shingles that are designed to last a minimum of 15 to 20 years, age alone is a primary reason that many homeowners must reroof.

Home builders, roofing contractors, and homeowners often choose asphalt-fiberglass shingles because of the initial low cost of the materials, the widespread availability and variety of asphalt-based products, and the ease of application. However, the lighter weight and relatively lower durability of asphalt shingles compared with a cement, clay, metal, slate, or composite roofing material means that asphalt-fiberglass shingles are more susceptible to wear and damage from the elements.

Until they detect a roof leak, most homeowners won't find it necessary to make more than periodic, cursory inspections of their roofs. Older roofs shingled with asphalt-fiberglass products can leak from storm damage or from loose flashing around a chimney or vent. Other indications that roof materials soon will need to be repaired or replaced are missing shingle tabs or buckled or warped and aged shingles.

If you live in a suburban development or an area where all the houses were constructed within a year or so of each other, and quite a few of your neighbors have been reroofing their homes, you can reasonably deduce that your roof also is ready for replacement. Nevertheless, don't be rushed into a hasty decision by the actions of neighbors. Inspect your roof carefully before deciding to make repairs or to completely reroof your home.

Salespeople for roofing contractors will often work an entire neighborhood and attempt to convince potential customers that every roof in the area is worn. Some salespeople will even strongly suggest that all old shingles be torn off before a new layer of shingles is installed. Actually, old shingles must be torn off only if they are buckled or warped or if the roof already has two layers of shingles. Whether you do the work yourself or have a contractor do the job, tearing off shingles greatly increases the time, effort, and cost to complete a roofing job.

If your shingled roof is less than 15 years old and leaks are apparent, minor repairs will probably solve the problem. If the shingles on your home were installed 15 to 20 or more years ago, inspect the condition of your roof very carefully.

Inspection Procedures

The cause of an asphalt-shingled roof leak is not always easy to find. It is not unusual for water to seep through cracked or worn flashing along a wall, chimney, or valley, and show up on interior walls as water stains many feet from where the water first entered the roof. Periodic, cursory inspections over the years you own your home can usually help you to spot signs of roof trouble. In many cases, binoculars can be used to make inspections without having to climb to the roof.

Binoculars are especially useful for inspecting potential problems from the ground with white or gray shingles, but it is harder to identify trouble areas on dark-colored shingles or black shingles. If your home has interior water damage, be prepared to climb on the roof to determine the extent of the trouble and to closely inspect the shingles for less than obvious but just as serious signs of worn materials. If you decide to make a close inspection of the surface of the roof, keep in mind that merely walking on worn and badly aged shingles can potentially add to the damage.

To determine whether your roof will need repairs or a complete reroofing, inspect the roof surface with the following points in mind. If you find that your roof fits the descriptions in either or both of the last two categories, your home definitely needs new shingles.

Missing tabs or shingles   Look for missing tabs or shingles. A few shingles or tabs blown off during a storm can usually be replaced, but if a roof section is missing many tabs or entire shingles and the damage is not from a recent storm, you probably have a worn roof.

Roof intrusions   Look around skylights, vents, pipes, and valleys for cracked or worn metal, aged caulking, and deteriorated roof cement. Intrusions in roofs are often the most frequently weather-stressed areas and, therefore, the most common locations where leaks originate. If the leak is not serious, such problem areas usually are not difficult to repair with roof cement and, if necessary, roof cement combined with some fiberglass mesh.

Flashing   Check the metal flashing around chimney walls, valleys, and dormers. Look for cracked, worn metal and deteriorated roofing cement and loose nails or loose step flashing. Leaks around flashing are sometimes difficult to locate and fix.

Fungus   If you discover a green mold or mildew on shaded areas of your shingled roof, you could remove the fungus by spraying the area with a mixture of detergent and pressurized water. The cure, however, might become a drawback. Instead of improving the appearance of your roof, you might dislodge a large amount of shingle-protecting granules. See chapter 14 for additional information.

General condition of shingles   Worn shingles (Fig. 4-1) have tabs that are cracked or curled (Fig. 4-2) at the edges and a significant loss of color and granules will be apparent. Large amounts of dislodged stony surface granules will appear in gutters and at downspout outlets where they have been washed off the shingles. When the topcoat of granules wears off a shingle tab, patches of asphalt can be seen. Apply pressure to a piece of a shingle to determine if it easily breaks or crumbles.

Worn shingles   Look closely between shingle tabs to determine if many granules have been dislodged. If the shingles are severely aged, you'll be able to see the wood deck through very small holes that have worn through the shingles. If there are many similarly worn shingles, it is a sure sign that your roof needs to be reshingled.

If your home has an attic or crawlspace and the roof has plank decking instead of plywood, you might be able to detect the same small holes from the inside of the home. The holes will appear as points of daylight shining through the worn shingles into the crawlspace.

Asphalt- and Fiberglass-based Shingles

Asphalt fiberglass-based shingles are produced in many brand names, colors, styles, textures, thicknesses, dimensions (including metric), and weights and continue to be the roofing material most commonly available through local suppliers. Some manufacturers offer asphalt fiberglass-based strip shingles designed with a layered texture resembling wood shakes, shingles with as few as two or as many as five tabs, shingles with two or three cutouts, or strip shingles with no cutouts. Some diamond, hexagonal, and T-shaped shingles also are available for *high-wind areas.* See Table 1-1 for a comparison of common asphalt-fiberglass shingle styles, weights, and dimensions.

Although asphalt fiberglass-based shingles are susceptible to cracking if they are installed during cold weather, the light weight of the fiberglass allows for a lighter shingle coupled with a longer limited-warranty period. The fiberglass content also allows for greater fire resistance than that provided by asphalt shingles.

Asphalt shingles, or so-called *composition shingles* or *organic shingles*, are made with layers of heavy roofing felt (made from organic mate-

Table 1-1 Styles, Weights, and Dimensions of Roofing Materials.

Material	Configuration	Per Square			Size		Exposure in inches	Specifi-cations
		Weight in pounds	Shingles	Bundles	Widh in inches	Length in inches		

Fiberglass-shingles

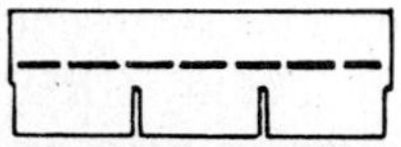	3 tabs	210 or 215	78 or 80	3	12 or 12¼	36	5 or 5⅛	UL Class A

Organic, three-tab shingles

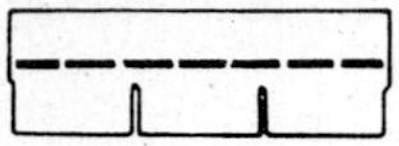	3 tabs	235 to 290	78 or 80	3 or 4	12 or 12¼	36	5 or 5⅛	UL Class C

No-cutout shingles

	various	235 or 240	78	3	12¼	36	5	UL 55B

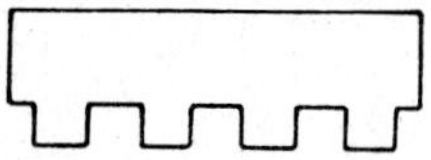	tabs give wood-shake look	250 or 300	100	4	12	36	4	UL 55B

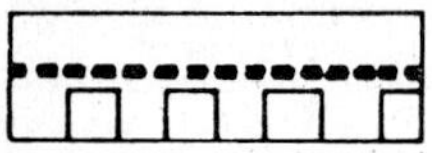

		AMERICAN STANDARD						
	two-ply strip shingles give layered look	320	84	4 or 5	11½	36	4¾	UL Class A
		METRIC						
		320	66	4 or 5	13¼	39⅜	5⅝	

rials such as paper or wood chips) that are saturated with asphalt. A thick coat of asphalt and minerals is then added to the saturated felt. A layer of *ceramic granules* or opaque-rock mineral granules are then added for color and for weather and sunlight resistance. In addition, *seal-down strips* of adhesive are added to the face (or in the case of at least one major manufacturer of shingles—Elk Corporation—to the back) of each shingle. After the shingles have been properly installed on the roof, the heat from sunlight activates the adhesive to seal down each shingle.

Table 1-1 Continued.

Material	Weight		Square feet per package	Length in feet	Width in inches	End lap in inches	Top	Exposure	Specifications
	Per roll	Per square							
Mineral-surface roll roofing	75 to 90	75 to 90	About one	36 or 38	36	6	2 or 4	34 or 32	
Double coverage, mineral-surface	55 to 70	55 to 70	one-half	36	36	6	19	17	
Coated roll roofing	43 to 65	43 to 65	one	36	36	6	2	34	43, 55, or 57 pound / 65 pound
Felt	52 / 60	15 / 30	4 / 2	144 / 72	36 / 36	4 to 6	2	34	15 pound / 30 pound

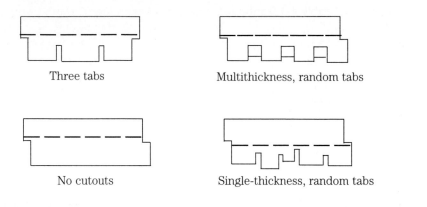

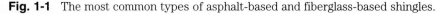

Three tabs   Multithickness, random tabs

No cutouts   Single-thickness, random tabs

Fig. 1-1 The most common types of asphalt-based and fiberglass-based shingles.

While a few shingle manufacturers still produce some asphalt-based composition shingles, composition shingles have been supplanted in the market by asphalt fiberglass-based shingles. Asphalt fiberglass-based shingles are manufactured using a process quite similar to that for asphalt shingles. Essentially, the difference is the use of an inorganic fiberglass base that is saturated with asphalt. Fiberglass-based three-tab shingles also feature self-sealing, sunlight-activated strips of adhesive applied during the manufacturing process.

Shingle Weights and Life Spans

Asphalt shingles are manufactured in a wide variety of colors, styles, and weights. Because of the comparatively low initial cost and ease of application, three-tab, 20-year, asphalt fiberglass-based shingles (Fig. 1-1) are by far the most common roofing material used on recently constructed or recently reroofed homes.

An important way to categorize roofing shingles is by weight per 100 square feet. Roofers commonly refer to 100 square feet of roof *coverage* as a *square*. Until the introduction of shingles made with fiberglass, it was accurate to presume that greater shingle weight meant longer life for the shingle. However, the combination of fiberglass and asphalt in shingles now permits lighter weight to be combined with longer shingle life.

The lightest, three-tab, asphalt fiberglass-based shingles weigh 225 pounds per 100 square feet, have a 20-year limited warranty, and are classified by Underwriters Laboratories as class A. *Class A* is the highest possible fire-resistance rating for shingles. The lightest, three-tab, asphalt shingles weigh 235 pounds per 100 square feet, have a 15-year limited warranty, and are classified by Underwriters Laboratories as class C. *Class C* is the fire-resistance rating for most conventional asphalt shingles.

Most manufacturers of asphalt fiberglass-based shingles offer shingles in 225, 260, 300, and 325 pounds per square. As the weight per square increases, so does the shingle life span and, not coincidentally, the purchase price for the materials. The cost of labor charged by contractors for installing the various weights of shingles usually does not change with shingle weight per square, but a steeply pitched roof to be covered with the heaviest-per-square shingles might increase the price of any contractor's bid. The lightest-weight shingles are designed to last 15 to 20 years, the middleweight range of shingles is designed to last 25 to 30 years, and the heaviest shingles are made to weather 30 years or more.

Several manufacturers have introduced layered fiberglass-based asphalt shingles designed to provide a textured, wood-shingled, or slate look. When installed, these distinctive *architectural shingles*, or so-called *laminated asphalt-fiberglass shingles*, provide a shadow line or random edge. Featuring dense, double-layer, laminated construction, architectural shingles often have a 35-year or 40-year warranty.

Many building suppliers are familiar with and will stock only the most popular asphalt-fiberglass shingle weights and styles, so be prepared to shop around with several suppliers in order to see different products. If you aren't able to find local distributors or wholesalers for the products you want, consult the Resources list in this book and ask for the dealers nearest your location.

As a general comparison that will vary with local conditions, you can expect wood shingles to last 10 to 40 years; clay, cement, or ceramic to last 20 to 60 years; metal to last 15 to 40 years or more; and slate to last 75 to 150 or more years. Prices for non-asphalt-fiberglass-shingle materials as well as contractor installation costs generally will rise with the increased durability and the increased difficulty of installation.

Shingle Color

A wide variety of shingle colors is available. White, off-white, light gray, or very light pastel colors on your roof reflect more sunlight and ultraviolet light during the summer than black or dark shades of asphalt-fiberglass shingles. Conversely, light colors absorb less sunlight and ultraviolet light during winter than dark shingles. Nevertheless, adequately combined soffit and attic ventilation are significantly more important factors in lowering the cost of cooling or heating your home than shingle color. Other factors, such as the amount of shade, orientation of the house toward the sun, exposure to wind, attic ventilation, and the pitch of the roof, all affect how long shingles will last.

Because dark-colored shingles absorb more heat during the day than light shingles, they are subjected to a wider range of contraction and expansion over their life spans. The result could be that black shingles might fail a year or two sooner than white shingles installed on a similar roof.

During very warm weather, any color shingle will mar or scuff from foot traffic. Black shingles are likely to scuff sooner because they gain heat faster. Also in hot weather, black shingles are more difficult for the applicator to install because individual shingles quickly become warm to the touch and installed courses are too hot to sit on in the proper shingler's position.

In cool weather, white asphalt-fiberglass shingles can become more brittle than dark shingles and, therefore, will be slightly more inclined to crack as they are installed. During winter, any warmth gained in the living space or heat lost from your home because of shingle color will be too small to measure.

If either summer cooling or winter heating is a very important factor at your location, select an appropriate color. Otherwise, choose a shingle color that complements the color scheme of the entire house.

Roof Pitch

The *pitch* or *slope* of a roof (Fig. 1-2) is very much a factor in the life span of asphalt-fiberglass shingles. Generally, the steeper the angle of the roof, the longer the shingles will last. Suppose that you own a Cape Cod-style house and your next-door neighbor has a ranch house. If the same brand and weight of shingles are applied to both houses, the shingles on the ranch house will wear out several years before the shingles on the Cape Cod house. Because of the steeper pitch of the roof on the Cape Cod house, the shingles will be exposed to less direct sunlight each day. In addition, rain running off the steeper slopes will generate less drag and, therefore, fewer granules will be dislodged.

The angle and intensity of the wind, rain, and especially sunlight as they strike a roof differs with the position of the home. The amount of sunlight and wind and the frequency of freeze/thaw cycles a roof is exposed to relate directly to the number of years an asphalt fiberglass-based shingle roof will last. Even the shade from trees or the wind deflection provided by a hill can make a difference (Fig. 1-3) in the life span of shingles. It is not unusual for the "sun side" of a roof to wear out several years before the "backside" of the same roof.

Other Materials

About 20 percent of American homes are roofed with materials other than asphalt-fiberglass shingles. While asphalt-fiberglass shingles are often considered to be the easiest roofing material to work with, one of the lightest roofing materials in weight per 100 square feet of coverage, and have the lowest initial costs when minimum-weight shingles are used, none but the heaviest, thickest, and most expensive laminated fiberglass-asphalt shingles approach the durability and long life of cement, clay, metal, slate, thermoplastic resin, or, in some climates, treated wood. As with asphalt-fiberglass roofing, the durability of wood shingles and shakes varies considerably with the thickness of the product, the local climate, and the design and orientation of the house.

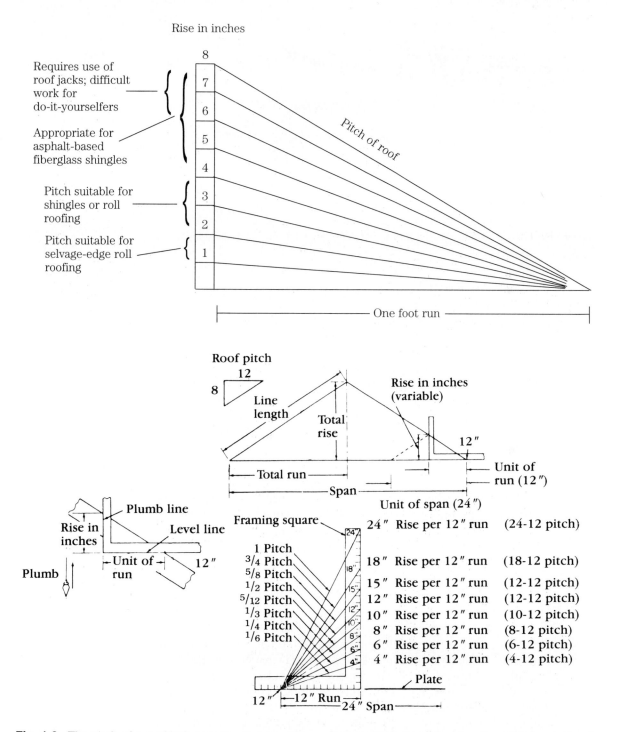

Rise in inches

Requires use of roof jacks; difficult work for do-it-yourselfers

Appropriate for asphalt-based fiberglass shingles

Pitch suitable for shingles or roll roofing

Pitch suitable for selvage-edge roll roofing

Pitch of roof

One foot run

Roof pitch

Line length

Total rise

Total run

Span

Rise in inches (variable)

12″

Unit of run (12″)

Unit of span (24″)

Plumb line

Level line

Rise in inches

Unit of run

12″

Plumb

Framing square

1 Pitch
3/4 Pitch
5/8 Pitch
1/2 Pitch
5/12 Pitch
1/3 Pitch
1/4 Pitch
1/6 Pitch

24″ Rise per 12″ run   (24-12 pitch)
18″ Rise per 12″ run   (18-12 pitch)
15″ Rise per 12″ run   (12-12 pitch)
12″ Rise per 12″ run   (12-12 pitch)
10″ Rise per 12″ run   (10-12 pitch)
8″ Rise per 12″ run   (8-12 pitch)
6″ Rise per 12″ run   (6-12 pitch)
4″ Rise per 12″ run   (4-12 pitch)

Plate

12″   12″ Run
24″ Span

Fig. 1-2 The pitch of a roof indicates the degree of the roof slant. Stated as a fraction, the pitch is a ratio of the rise to the span of a roof. Most homes have roofs that rise between 2 and 5 inches per foot.

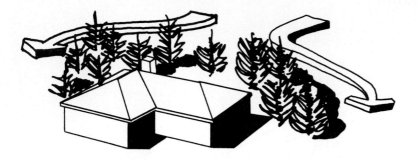

Fig. 1-3 The wind deflection and the shade created by trees can add years to the life span of roofing materials. The roofing on the sunlight-exposed side of a house positioned like this will likely wear out several years before the shaded side.

The annual cost per 100 square feet for the most common roofing products can be estimated at approximately $8 for 15-to-20-year asphalt-fiberglass shingles; $5 for 50-year clay tiles; $5 for 50-year cement- and wood-fiber composites; $10 for 40-to-50-year galvanized metal; and $25 for 10-to-15-year (the thinnest) treated wood shakes. Contractors generally charge about $100 to $120 per square to install asphalt-fiberglass shingles; $130 to $150 per square to install concrete roofing; and $240 per square for unglazed clay tiles and over $1000 per 100 square feet for glazed tiles. Slate and wood roofing installation prices vary with the availability of contractors and products, but you can expect labor costs to be two or three times the price of installing asphalt-fiberglass shingles.

Cement

Cement, concrete, and cement- and wood-fiber composite products resemble sunbaked clay tiles or slates. Cement shakes are manufactured to imitate wood shakes when installed on a roof. If you can accept the above-average initial costs for materials, equipment and labor, cement-roofing products offer considerable advantages over other popular roofing choices. Cement roofing is durable in wet or dry climates and tile has a class A fire rating. In Europe and Australia, about 90 percent of the houses have some form of tile roofing.

With any cement roofing material, *efflorescence* (a chemical change resulting in incrustation or powdery substance) and hue variance can occur. The most popular colors for cement tiles are earth-tone browns, tans, and slates, but the addition of iron-oxide pigment to the mix can produce white, green, or even blue tile. Some manufacturers coat their tile with a slurry of cement and pigment to add color.

Cement products require a minimum of 30-pound felt as an underlayment. A layer of 45-pound felt, a double layer of 30-pound felt, or a combination of felt and roll roofing will greatly extend the life of the underlayment. Keep in mind that the underlayment—not the tile—waterproofs a cement-tile roof.

Cement tile roofs require a raised *fascia* and an *anti-ponding strip* of metal flashing at the eaves so that the first course of tiles is installed at the same pitch as the other courses. A watertight underlayment and horizontal *lathing* installed 14 inches apart must be nailed to the roof deck. Some tile installation techniques also require vertical *battens* and horizontal lathing. Refer to chapter 12 for guidelines on installing a typical cement-tile roof.

Composite roofing products, such as FireFree shakes, shingles, and slates (by Re-Con Building Products) and Cemwood shakes, slates, and tiles (by American Cemwood), combine Portland cement and wood fiber to provide a class A fire rating, a limited 50-year warranty, and a relatively lightweight roofing material. FireFree products weigh 375 to 395 pounds per 100 square feet and Cemwood shakes weigh 450 to 580 pounds per square feet. By comparison, concrete tiles weigh from about 900 to 1200 pounds per 100 square feet, and wood shakes weigh about 350 pounds per square.

Composite roofing is suitable for new homes as well as reroofing projects, and the shakes, shingles, and slates can be walked on once they are installed on a suitably sloped roof. Because composite materials are less brittle than clay, composites are less likely to break, chip, or crack during transport and installation.

Clay

Mission tiles are semicircular or barrel-shaped. *Spanish tiles* are S-shaped. *Greek tiles* are flat and are bridged by semihexagonal cap tiles. The orange-red colors of clay tile will not fade or produce efflorescence. The substantial weight of clay tiles requires strong roof framing. Clay tiles are expensive to purchase but will last 50 years or more. As with cement roofing, clay roofing products depend on the underlayment for waterproofing.

Be sure to install long-lasting roofing materials such as one 45-pound felt layer or two layers of 30-pound felt underlayment over mission, Spanish, or Greek tiles. Otherwise, the underlayment will fail long before the clay tiles themselves, and the tiles will have to be removed, new felt will have to be installed, and the tiles will have to be reinstalled on the roof deck.

Specific installation guidelines for clay tile roofs vary with each manufacturer's tile pattern and features, but generally, the roof deck is felted, chalk lines are laid as guidelines for courses, and—if the tiles are designed to be hung on battens—battens and laths are nailed to the roof deck.

A raised fascia board, special under-eaves tiles, or *birdstops* (for mission tiles) must be installed at the eaves before the first clay tile course is installed.

Natural clay products can vary in color. To avoid disappointment regarding color selection, inspect actual samples at the time of initial product and color selection and again at the time the product is delivered to the building site.

To avoid potential undesirable color patterns, the installer must periodically inspect the roof tiles from the ground at a distance of about 50 feet.

Periodic inspections from the ground will indicate the best way to obtain a random color sequence, ensure an acceptable blending of colors, and help avoid streaks, stair-stepping, or checker-boarding.

Metal

Metal shingles, shakes, and sheet-metal panels are manufactured from aluminum, stainless steel, galvanized steel, or copper to resemble wood shakes. With proper maintenance, metal roofing products can last 50 to 75 years. Note that nongalvanized *metal roofing* must be painted every few years.

Revere Copper Products offers lightweight—142 pounds per square—copper roofing panels for about $350 per 100 square feet. Check with local contractors regarding availability and installation estimates.

Refer to chapter 10 for additional information on metal roofing.

Photovoltaic Modules

More than 20 years ago, the Japan Photovoltaic Energy Association (JPEA), which is composed of more than 80 companies and organizations, began encouraging research on photovoltaic power generation. Today, solar energy products have entered a new phase of acceptance as worldwide photovoltaic products have reached a reported $1 billion a year in sales.

Sanyo Solar Energy System Co., Ltd., Kyocera Corporation, Kaneka Corporation, Kubota Corporation, Misawa Homes Co. Ltd., and other companies have developed or are working on solar-cell modules for the Japanese market that can be used as roofing materials. The goal of these companies is to develop essentially amorphous and polycrystalline silicon-based, solar-powered batteries that also serve as weather-resistant residential roofing materials in the form of modules or flat panels.

Until recently, little thought was given to developing roof-mounted solar-powered panels as a roofing material. The conventional method for installing solar devices is to attach a platform to the existing roof and hope that the weight of the device does not cause leaks in the roofing material. Other typical concerns included potential solar-cell damage by snow, ice, hail, or high winds or that the unit might become covered by leaves.

In July 1994, Misawa Homes introduced solar-battery modules—that function as the residential roof surface—to the Japanese housing market. The modules are weather resistant, highly durable, and do not require skilled labor to install.

Because Misawa's roof modules incorporate reinforced glass as a main component, the product will last—much like the glass in high-rise buildings—indefinitely. And this system is designed to feed back surplus power to the local utility. Misawa is awaiting formal authorization from Japan's Ministry of Construction to use the product as a roofing material.

Sanyo's Japanese advertisements for the solar-power-generated roofing system promote the advantages of clean power from sunshine, maintenance-free operations, easy interface with a conventional power supply as needed, and the capability of selling surplus power to the local power company.

Coincidentally, California, under a provision effective January 1996, permits residents to receive the full value of electricity they generate—up to approximately 15 cents per kilowatt-hour—and to make use of their existing electric meters to keep track of the electricity they produce. The new law also allows homeowners to generate as much as 10 kilowatts of electricity from a residential photovoltaic system.

With a photovoltaic system installed, meters will run backward when power is flowing from the homeowner's system onto the power company's grid. Under previous California law, a homeowner had to have a separate meter to track the flow of electricity from the home to the power company. In addition, power companies previously paid only 3 cents a kilowatt-hour to residents.

According to representatives of Southern California Edison and San Diego Gas and Electric Co., the market for photovoltaic roofing materials should be wide open. Of Edison's 4 million or so customers, fewer than a dozen now generate their own power. In San Diego County, no one has a photovoltaic system connected to the local power grid.

Polymer Resins

Very lightweight, thermoplastic resin and *polymer roofing* tiles and panels designed to look like shakes or slates have been introduced to the roofing product market by companies such as Nelco Engineering, Inc. and Everest Roofing Products.

Nelco slates or shingles do not have to be replaced in your lifetime, can be walked on, weigh an extremely lightweight 70 pounds per 100 square feet, and are made of thermoplastic resin panels that can be installed without battens on a new deck, where worn shingles have been removed, or over one layer of worn asphalt shingles.

Everest polymer tiles have a 50-year life, are very light at 110 pounds per 100 square feet, can be walked on, and require no base of battens or lathing when installed.

Slate

Made from quarried stone, *slate roofing* is most commonly used as an expensive option in the north and northeast regions of the country. For other areas of the country, transportation costs add substantially to the initial high cost of slate roofing. The weight of slate—750 to 850 pounds per square—requires strong roof framing, and depending on where it was quarried—Pennsylvania, Virginia, or Vermont—slate can last 75 to 150 years. Quarry prices generally range from $300 to $400 per 100 square feet.

Slate colors are dictated by the chemical and mineral content of the original clay sediment and include gray, blue-gray, blue-black, black, purple, mottled green and purple, and red. Ribbon slates are streaked with impurities from the original clay. Clear slates do not have streaks and, therefore, do not have the tendency of ribbon slates to weather prematurely where the streaks of impurities occur.

Slate roofing contractors are not common. Installing a slate roof is not considered a practical do-it-yourself project because of the substantial weight of slate, the special tools used, and the special skills needed to install it.

Recycled slate from a demolished building can be used if the slate is not very old (perhaps 30 or so years old), but the slates must be safely removed from the old roof, transported, and cleaned in a solution of oxalic acid and water. In the unlikely event that you are able to locate enough salvageable materials, it takes special skills and experience to complete the work. Finally, slate cannot be walked on because it breaks easily.

Wood

Wood roofing is made from red cedar, redwood, cypress, pine, or oak. *Wood shingles* are sawn by machine; *wood shakes* are hand split and are generally thicker or rougher looking than wood shingles. Red cedar has a natural oil that discourages splitting and encourages rainwater to be shed.

Wood roofs are designed so that the shingles or shakes absorb moisture, swell during the absorption process, and, as part of the swelling process, shed additional rainwater. Shingles and shakes contract as sunshine and air currents aid in evaporating water absorbed by the wood roof. Therefore, wood shingles and shakes are best suited for homes in locations with a varied, temperate climate.

Extended, intense sunlight or prolonged, complete shade will lower the life expectancy of wood shingles and shakes. Extensive sunlight dries and cracks wood; complete shade discourages evaporation and encourages moss, fungi, and wood rot.

Some communities restrict or ban wood roof shingles and shakes because of their inherent flammability. Treating wood shingles and shakes with chemicals to resist fire adds about 25 percent to the cost of the material. Homeowner insurance rates can be higher or might not be available for a home with a wood roof. Refer to chapter 9 for additional information on installing wood shingles and wood shakes.

Chapter **2**

Tools and Equipment

*H*aving the proper tools and equipment is a very important part of any roofing work. Some of the specialized items you'll need might be somewhat difficult to obtain locally. Even well-stocked hardware and home-improvement stores don't carry all types of roofing tools.

The types of tools and equipment described in this chapter can be ordered from the ABC Supply Company, Inc. catalog. ABC Supply Company is a national roofing, siding, and insulation products and materials distributor. See the Resources list in this book.

Hand Tools

As a minimum for shingling even the easiest of roofs, you'll need a roofer's hatchet or a carpenter's hammer, a nail pouch, a utility knife and blades, a chalk line and chalk, a tape measure, shoes with rubber soles, and a ladder. Other specialty tools and equipment are designed to make your roof work safer, easier, and more efficient.

Broom or Brush

For tear-off work, you'll need a broom to sweep away debris, shingle granules, and nails from the exposed roof deck (Fig. 4-24). A stiff-bristle brush can also be used to spread roofing cement for roll roofing.

Circular Saw

For making repairs to damaged plywood sheathing or roof-deck planking, you'll need a circular saw, one or two extension power cords, and access to electricity (see Fig. 2-1).

Fig. 2-1   For repairs to roof-deck planking and plywood, use a circular saw.

Claw Hammers or Ripping Hammers

Common hammers can be used for a limited amount of shingling, and claw hammers or ripping hammers are useful for removing or loosening vent flanges and pulling out old, stubborn flashing and nails. Nevertheless, a round, lightweight head (Fig. 2-2) is a disadvantage for driving roofing nails using the rapid, rhythmic technique favored by skilled roofers. Any do-it-yourself roofer who believes there is such a thing as a multipurpose hammer can expect bruised fingers and slow work.

The smooth surface of the head of a carpenter's hammer is designed for driving nails flat to the surface of wood without leaving a mar. Therefore, a carpenter's hammer is adequate but not ideal for nailing shingles. Don't use a tack hammer, a mallet, or a ball-peen hammer for shingling.

Caulking Gun

Use *caulk* (Fig. 2-3) to enhance waterproofing of flashing around a pipe flange or at a chimney where counterflashing is used. Do not use caulk as a substitute for roofing cement.

Chalk and Chalk Line

Blue, red, or yellow chalk is most often sold in plastic tube containers. The chalk, the chalk box, and the chalk line must be kept dry. When you fill a

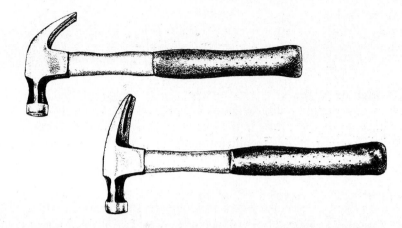

Fig. 2-2 A claw hammer (top) or a ripping hammer is a useful tool for prying flashing or nails. Either hammer can be used for limited roofing work, but a roofer's hatchet (Fig. 2-6) is more appropriate for nailing shingles.

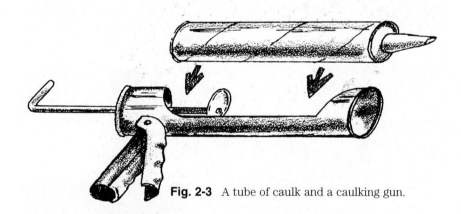

Fig. 2-3 A tube of caulk and a caulking gun.

chalk box, pull about half of the line out of the chalk box and pour in only enough chalk to fill the box halfway. If you overfill the box with chalk, the line will jam. Rewind the line and be careful to avoid causing knots or tangles. Pull out the line about 2 to 3 feet to see if the chalk has saturated the line. If the line hasn't been saturated, pull the line out halfway and add a little more chalk to the box.

When you are snapping lines on a roof, it is generally possible to snap four or five lines before you must rewind the line. It will not be necessary to refill the box with chalk each time you rewind the line.

Gloves

Cotton work gloves are essential for gripping your tools while tearing worn shingles from a roof. You also will find gloves useful when unloading bundles of new shingles from the delivery truck.

Hammer Tacker

Duo-Fast, Bostitch, or Arrow hammer tackers (Fig. 2-4) are ideal for use by professional roofers who must rapidly staple building paper. Such manually operated hammer tackers are especially useful for applying felt to mansards.

Nail Pouch or Apron

A nail pouch should be large enough to hold multiple handfuls of roofing nails. Making excessive trips to the nail box wastes time and effort. Larger pouches often have separate compartments that are convenient for holding a utility knife, a tape measure, and other small items.

Power Saw

A gas-powered saw for cutting cement roofing tiles costs about $800 and its blade costs about $300. Make sure such equipment is available for rent if you contemplate taking on a do-it-yourself tile-roofing project (Fig. 2-5).

Roofer's Hatchet

A roofer's hatchet is designed for professional roofers who typically expect to drive a 1-inch roofing nail with one stroke. If you plan to shingle by hand, buy a roofer's hatchet that feels comfortable in your grip. The square head

Fig. 2-4   Hammer tackers can be used to apply felt quickly.

Fig. 2-5 A gasoline-powered saw for cutting tile.

on a roofer's hatchet is large and heavy and the head has a gridded surface (Fig. 2-6). Some roofing hatchets have a sharp blade for splitting wood shakes. Other hatchets have a small knife blade for cutting asphalt-based shingles. It is much easier to cut yourself with the small, very sharp blade than it is to use the blade to cut shingles.

Safety Eyeglasses
Many hardware and home-improvement stores stock inexpensive, plastic safety goggles, face shields, and other types of safety glasses. Some shields are designed to fit over the top of standard prescription eyewear. Regular prescription glasses and contact lenses do not provide protection from nails, staples, or pieces of wood or metal thrown up by a roofer's hatchet or by power tools. Eyewear that meets American National Standards Institute requirements are marked with a "Z87."

Shoes
Don't underestimate the importance of wearing sturdy but comfortable rubber-soled shoes while you are working on your roof. Other types of shoes will not give you the proper footing you need to work on a sloped roof. Proper shoes with superior traction are of utmost importance for working on steep roofs. Don't wear leather-soled shoes on a roof.

The milled face helps prevent flying nails

Fig. 2-6 A shingling hatchet has a large, square head with a milled face. The adjustable gauges provide the roofer with a handy means of checking the alignment and exposure of a shingle as it is positioned and nailed.

Another factor in selecting proper footwear is that hard-soled shoes will quickly mar newly installed shingles. Check to see if your shoes have large, metal hooks for securing the laces. Use tin snips to cut off such hooks. If you do not remove the hooks, you will continually mar the shingles by dragging your feet over the shingles as you work in the proper roofer's sitting position.

Shovels, Roof Rippers, and Tear-Off Tools

A flat-bladed, tear-off shovel (Fig. 2-7 and Fig. 2-8) is the only type of shovel to use when you must remove worn layers of asphalt shingles or roll roofing. Don't try to use a garden shovel to tear off roof shingles.

Many professional roofers prefer to use a tear-off shovel that has a spot-welded "block," or fulcrum, on the back (Fig. 2-8) for leverage when pulling nails. Keep in mind that shovels with notched blades will catch and help pry nail heads. Straight shovel blades work well on roll roofing. Notched or straight blades are suitable for tearing off asphalt shingles.

For tearing off wood shingles, a spade that resembles a pitchfork, for getting under and lifting the wood shingles, and a wide-base shovel, for scooping up debris, is helpful.

For large projects, Aaron makes a wheeled, 42-pound Super Tear Off Bar—designed to be run under courses of old roofing material—for about $210 (replacement blades cost about $65 to $80). The ABC Supply Company catalog lists a wide range of various manufacturers' specialty tools for roofing tear-off work.

Tape Measure

A tape measure doesn't have to be elaborate but it does have to be accurate. A 50-foot tape is convenient for measuring the total square feet of a roof. A 6- or 8-foot tape will do the job of the longer, more expensive tape if you are very careful when you take measurements.

Fig. 2-7 Use a flat-edge shovel for tearing off asphalt shingles.

Tin Snips or Aviation Snips

Along with cutting metal for flashing and valleys, *tin snips*, or so-called *aviation snips* (Fig. 2-9), can be used to cut shingles. For the inexperienced roofer, cutting shingles with a utility knife is often a difficult and time-consuming task to learn. To avoid scraped knuckles or worse injuries to your hands, snap a chalk line to obtain a straight line and cut the roofing material with tin snips. Ask your tool supplier for snips designed for either right-hand or left-hand use.

Fig. 2-8 Tear-off shovels with a welded block (right) provide added leverage for removing worn shingles.

Fig. 2-9   When you purchase aviation snips or so-called tin snips, look for a pair designed specifically for a left-handed or a right-handed person, or for straight-ahead cutting. All three types of snips are manufactured.

Trowel

A trowel (Fig. 2-10) makes easy work of applying roofing cement. Use the flat edge of the trowel to smooth out an even layer of roofing cement around a vent or chimney.

Fig. 2-10   A trowel for applying roofing cement.

Utility Knife and Blades

The roofing tool that is the most difficult to learn to use properly and skillfully is the utility knife (Fig. 2-11). *Straight blades*—used primarily by carpenters—are easy to find in stores, but *hook blades* are the easiest and safest type of utility blades to use for cutting shingles. Some roofers, however, prefer to use a larger, more expensive knife designed to cut linoleum. Make sure that the blade is properly secured in the knife you choose. Be careful!

Additional Equipment

For complicated roofing work, some additional tools will be needed. Use the following descriptions of equipment as a guideline on determining the tools and equipment required for the type of building on which you will be working and the types of materials you will be applying on your roof.

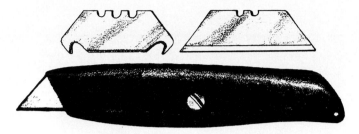

Fig. 2-11   A utility knife, a hook blade, and a straight blade. Hook blades are better for cutting asphalt or fiberglass shingles.

Roof Jacks or Angle Brackets

If the pitch of your roof is too steep for you to walk on comfortably, use *roof jacks* (Figs. 2-12 and 2-13) and planking as a safety precaution. Some roofs are so steep that roofing without jacks is impossible. If your roof seems even a little too steep to walk on comfortably, you might want to use jacks and planking to hold the bundles of shingles in place after you bring the

Fig. 2-12   On a steeply pitched roof surface, roof jacks and planks must be used.

Fig. 2-13 Roof jacks can be adjusted to accommodate the steepness of the roof pitch.

bundle up from the ground. *Angle roof brackets* cost about $6 to $8 and three-position adjustable angle brackets cost about $10.

Always use spikes (3-inch, 10d nails) to secure roof jacks. The jacks are designed so that they can be easily unfastened from the roof deck. When it is time to move the jacks and planking to a different part of the roof, it is much easier to drive the spikes through the shingle and into the deck than trying to pull the nails free.

Staple Guns, Nail Guns, Compressor, and Hoses

A pneumatic staple gun (Fig. 2-14) or pneumatic nail gun (Fig. 2-15), an air compressor (Fig. 2-16), and hoses are much too expensive to purchase for limited use. A new compressor, two new staple guns, and new hoses can cost $1000. Renting power tools can save considerable time and labor if your roofing job, particularly something such as multiunit apartment buildings, is large enough to justify using such equipment. A staple gun or nail gun will double the amount of shingles you can install in one day. A few telephone calls to check on the availability of rental equipment and costs could easily prove worthwhile for a do-it-yourselfer shingling a typical 20-square-foot roof.

Bostitch, Duo-Fast, Senco, Hitachi, and other manufacturers' coil nailers and staplers are specifically designed for applying asphalt or fiberglass roofing shingles. The lightweight (approximately 6 pounds) roofing *coil nailers* can drive up to 16-gauge wire, $1\frac{1}{2}$-inch electrogalvanized roofing nails. Use 1-inch nails for new work and $1\frac{1}{4}$-inch nails for roofing over the top of one layer of old shingles. Make sure that the nails or staples are driven flush with the shingle surface but not through the shingle or less than flush.

Be absolutely certain that the equipment you buy or rent is capable of driving galvanized or aluminum roofing nails or staples that are 1-inch long for one layer of shingles and $1\frac{1}{4}$-inches long for roofing over the top of a layer of old shingles. Never attempt to use brads to install shingles.

Fig. 2-14 Shingling with a pneumatic roofing stapler.

Because many roof failures were blamed on stapled shingles torn loose by sustained winds of almost 150 miles per hour and gusts to 200 miles per hour from 1992's Hurricane Andrew, the Southern Building Code International decided in early 1993 and the Asphalt Roofing Manufacturers Association (ARMA) decided in 1994 to list nails as the preferred method for fastening asphalt shingles to sheathing. In "high-wind" areas, ARMA also now recommends applying six nails per shingle. Note that these are recommendations only. Neither the code nor the guide-line bans the use of staples or requires the use of more than the standard four fasteners per shingle.

Laddeveyor or Material Hoist

A *laddeveyor* is a combination ladder and conveyor belt. An aluminum lad-der with a gas-driven motor (usually $1^1/_2$ to 4 horsepower), a pulley, a sled, and a bundle catcher are designed to lift bundles of shingles to one- or two-story roofs (Fig. 2-17). One laddeveyor is made by the Louisville Ladder Division of the Emerson Electric Company and another model, by the Safety Hoist Company, is available through the ABC Supply Company cat-

Fig. 2-15 A pneumatic nail gun can be used to nail felt as well as shingles.

alog. At a cost of about $1400 to $1700, a gas-powered material hoist is much too expensive to purchase for small jobs but it is definitely worth renting if you are not able to arrange for your shingles to be delivered to the roof surface by the supplier.

Many shingle suppliers have shingle-delivery equipment. Call the suppliers in your area to find a dealer with such equipment (see Figs. 3-4 and 3-5). The savings in labor will more than offset any additional, usually nominal, charge for delivery of the shingles to the roof.

Ladder

It is essential that the ladder you select be sturdy and of the proper length for the job. The four common grades of wood, aluminum, or fiberglass extension ladders are light duty (household use), medium duty (commercial grade), heavy duty (industrial use), and extra-heavy duty.

A classification and weight limit label found on the side of a ladder describes the safe working loads for each category of ladder. The duty rating includes a fully clothed user's weight and the tools and materials carried on the ladder:

Fig. 2-16 An air compressor for powering pneumatic nail guns.

Grade	Color Code	Duty Rating (lbs)
household	red	200
commercial	yellow	225
industrial	blue	250
extra-heavy industrial	black	300

If your shingles will be delivered to the roof surface, a medium-duty ladder—designed to support 225 pounds on a rung—is adequate for most homeowners. If your weight is above average and you plan to carry the shingles on your shoulders and up the ladder to the roof, use a heavy-duty or extra-heavy-duty ladder. Remember that most shingle bundles weigh at least 75 pounds.

Also, keep in mind that extension ladders are not designed to extend to full length. Because the ladder should extend 3 feet above the eaves,

Fig. 2-17 A laddeveyor.

only 21 feet of a 24-foot ladder can be used. Extension ladders are typically sold in 4-foot-length increments from 16 to 40 feet. The owner of a single-story home should select a 16- or 20-foot ladder. In general, a 24- or 28-
foot ladder is adequate for a two-story home.

Other ladder safety features to look for include a sturdy mechanism for locking together the two sections, nonslip anchor shoes, a reinforced bottom rung, and skid-resistant rungs. If your ladder doesn't have skid-resistant rungs, you can apply self-adhesive, nonskid strips to the rungs.

Along with a wide selection of conventional wood, aluminum, and fiberglass ladders on the market, several types of specialty products are available. Versatile scaffolds, stages, and articulated ladders are designed for use in awkward spaces—by employing hinges and folding sections—or to be converted to work platforms. Specialty devices will prove especially useful if you must shingle a mansard roof.

If your roof has sections where working directly from a ladder (Figs. 5-37 and 5-38) is the best way to install the shingles, a U-shaped stabilizer device—available for about $20 to $60—can be quickly attached to a ladder. The stabilizer can be used to free the work space around gutters and to span such obstacles as windows. A ladder hook, that costs about $20, can be used to clamp the top rung of a ladder over a roof's ridge line.

While wood ladders are somewhat less likely to mar siding and gutters and are less expensive than aluminum or fiberglass ladders, they are heavier and more difficult to maneuver. A typical 24-foot wood ladder weighs about 65 pounds, and a comparable aluminum ladder weighs 50 pounds. Fiberglass ladders are the most expensive of the three choices.

If you are using a wood ladder that is several years old, be absolutely certain that the ladder rungs will not break when you place your weight—plus the weight of a bundle of shingles on your shoulder—on the ladder (see Figs. 2-18 and 2-19).

Before you carry any shingles to the roof surface, read the section on bundle-stacking methods described in chapter 3, and keep the following safety guidelines in mind any time you use a ladder.

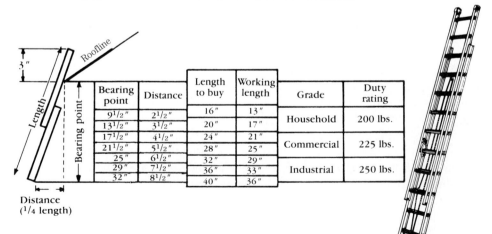

Bearing point	Distance	Length to buy	Working length	Grade	Duty rating
$9^1/2''$	$2^1/2''$	16″	13″	Household	200 lbs.
$13^1/2''$	$3^1/2''$	20″	17″		
$17^1/2''$	$4^1/2''$	24″	21″	Commercial	225 lbs.
$21^1/2''$	$5^1/2''$	28″	25″		
25″	$6^1/2''$	32″	29″	Industrial	250 lbs.
29″	$7^1/2''$	36″	33″		
32″	$8^1/2''$	40″	36″		

Fig. 2-18 Ladder guidelines.

- In the United States, according to insurance-industry statistics, almost 400 people are killed each year and some 40,000 individuals suffer injuries each year from ladder accidents.

- Remember that a ladder can conduct electricity. Stay away from all wires.

- Avoid leaning a ladder against vinyl or aluminum gutters and spouting, a window, or an unbarricaded, unlocked door.

- The base of the ladder must rest on an even surface and be placed so that the distance from the eaves is equal to one-quarter the length of the ladder.

- The ladder should extend about 3 feet above the roof surface.

- Never attempt to use a damaged ladder. Having the rungs of a ladder snap while you are halfway up a two-story climb is not a pleasant experience.

- If you are using an extension ladder, make certain that the locks are fully engaged.

- Wear rubber-soled, slip-resistant shoes.

- While on a ladder, don't attempt to stretch or reach just a little too far.

- Climb up and down the ladder facing the building and use at least one hand for balance and support. Tools can be carried on a tool belt or in a bucket.

- When you finish work for the day, take down the ladder and put it out of the reach of children.

Nail Bar or Pry Bar

A nail bar or pry bar (Fig. 2-20) is very handy for removing hard-to-get-at nails and flashing that must be pulled when you are tearing off shingles.

Hip Pad

If you are going to do a great deal of shingling, a rubber *hip pad* will save wear and tear on you and your jeans. As you sit on the roof, the pad straps snugly around the area of your body that drags across the abrasive surface of the newly installed shingles. A new pair of jeans can be worn out in a matter of days when subjected to this type of treatment. A hip pad will last for years, and it also provides some slip resistance and some insulation from shingles that absorb considerable heat from sunlight.

The McGuire-Nicholas Mfg. Co. is one source for hip pads that sell for about $30. When ordering such equipment, be sure to specify whether you are a right-handed or left-handed roofer. For a right-handed person, the pad will be strapped around the left leg; the opposite is true for left-handed roofers.

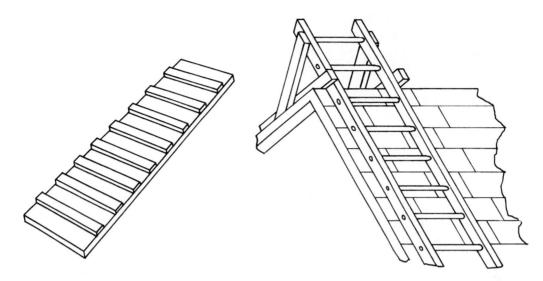

Fig. 2-19 A "chicken ladder" or a ladder with ridge hooks can be used on steeply pitched roofs.

Fig. 2-20 A pry bar.

Magnetic Nail Sweepers and Rakes

Roofing tear-off work inevitably sends many nails or staples to the lawn, driveway, parking area, or street below. Several magnetized tools—ranging in price from about $30 for a hand-held rake to $140 for a wheeled sweeper—will help you avoid potential injuries or tire damage. The simplest device, a magnetic bar, attaches easily to a spring rake.

Fall-Prevention Gear

Roof anchors, 16d nails, nylon rope, a locking carabiner (which resembles a giant clip-on safety pin), a shock-absorbing mechanism, and a body harness will allow you to work on all flat and pitched roofs within Occupational Safety and Health Administration (OSHA) standards. The minimal cost for such equipment ranges from about $400 to $500. After a while, you will become accustomed to not tripping over the nylon rope.

Polyethylene Tarps and Plastic Sheeting

Plastic tarps and sheeting can be used to protect grounds from spills, dirt, nails, and debris, as well as for temporary rooftop protection. Prices vary according to the size of the sheet or tarp. An 8-×-100-foot sheet costs about $30 and a 50-×-100-foot tarp with aluminum grommets every 3 feet costs about $260.

Chapter **3**

Preparations

Before you begin nailing shingles or before you decide to hire someone to do the work, your first task is to become an educated consumer so that you are prepared when you talk to potential suppliers and contractors. Make informed decisions before you decide to do the work or have a contractor do some or all the work.

By following the guidelines set forth in this chapter and by reading the information provided by the manufacturer of the roofing materials you purchase for your home, the general techniques described in this chapter can be applied to many styles and shapes of shingles and roofing materials. Be certain that you refer to the backs of your material wrappers or shingle bundles for specific manufacturer's instructions. Take special note of the exact dimensions and exposures specified for the brand you decide to order from the supplier. Shingle dimensions and specifications vary slightly among manufacturers.

Examine each section of your roof to determine if the pitch is too steep or if you feel you will be too high off the ground to work with confidence. Now is the time to walk over the roof surface to decide if you can do the work or if you should call professional roofing contractors for estimates.

Choosing a Contractor

After you have examined the condition of your roof and you decide that you do not want to do the work yourself, take the time to select a qualified contractor. A *general contractor* or contractor is someone you independently hire to do work on your home; a *subcontractor* is hired by the general contractor or contractor and should be covered by the general contractor's or contractor's insurance policies.

Licensing of general contractors is not required by the following states: Colorado, Idaho, Illinois, Indiana, Kansas, Kentucky, Maine, Missouri, New Hampshire, New York, Ohio, Oklahoma, Pennsylvania, South Dakota, Texas, Vermont, Wisconsin, and Wyoming. Texas does not require contractors to have workers' compensation coverage, most states exempt sole proprietorships from requirements to carry workers' compensation liability coverage, and some states exempt employers with fewer than three or five employees from requirements for workers' compensation liability insurance.

Generally, misunderstandings and disputes most frequently arise between contractors and homeowners regarding details not discussed before a contract is signed. Here is a checklist of topics to discuss with the salesperson before you sign a contract for roofing work on your home:

Money

- Free written estimate detailing the contractor's cost to the homeowner for specific work.
- Homeowner's payment provisions following final inspection.
- Consider specifying that the final payment to the contractor be dependent on your receipt of a release of *mechanic's lien* from all suppliers and subcontractors.
- How unforeseen expenses will be handled, such as repairs to broken sheathing discovered during tear-off work. Have such costs been included in the estimate?
- Authorization and payment for allowable *change orders* to the contracted work. Examples include substituting the brand, style, or weight of shingles and estimating the amount of materials needed.
- Obtaining and paying for a building permit from the local government.

Fine Print

- Setting the target dates for starting and completing the work and any provisions for acceptable delays.
- A list, as part of the proposal and final contract, of all materials to be used, including brand, style, weight, color of roofing and any items such as replacement sheathing, skylights, ventilators, and fasteners (staples or nails), and flashing.
- A list of the contractor's previous customers who will allow you to see finished work or work in progress.

Insurance

- Copy of proof of liability insurance.
- Copy of proof of workers' compensation insurance.
- Is the contractor a *bonded contractor* with insurance for major liabilities?

Warranties

- Contractor's warranty on work.
- Roofing material manufacturer's product warranty and registration form.

Local, County, and State Government

- Compliance with building codes and work permits.
- Compliance with local, county, and state business-license requirements.
- Does your state have a "recovery fund" designed to cover losses to homeowners for damages caused by failure of a licensed contractor to adequately improve the property?

Access to the Job Site

- Contact persons and telephone numbers for each party (homeowner and contractor) to advise of potential and actual schedule changes.
- Job-site access, if necessary, to electricity for contractor's power saws or electric nailers.
- Job-site access by material suppliers, and the contractor name and telephone number for the supplier's dispatcher.

Debris

- How recycling or disposing of debris will be handled, particularly for tear-off jobs.
- The type of job-site protection afforded: tarps for driveways, shrubs, flower beds, etc., particularly for tear-off jobs.
- How debris from gutters will be cleaned.
- How the final cleanup of the job-site will be handled.

Inspections

- Does the contractor employ an inspector to review the work in progress?
- How the final inspection of the work will be handled.

Callbacks

- Provisions for follow-up or callback procedures for correction of potential work deficiencies should be included.

The National Roofing Contractors Association, the National Association of the Remodeling Industry, and other experts suggest that you undertake your search for a contractor using the following guidelines.

- Ask neighbors, friends, and relatives for the names of recommended companies. Any prospective contractor or remodeler should be willing to supply references from jobs underway or work recently completed. You can telephone the National Roofing Contractors Association at (800) USA-ROOF for a computerized referral service for your area's association members.

- Ensure that the contractor has a permanent business address and telephone. Avoid fly-by-night contractors by checking with your local Better Business Bureau to determine if there is a history of dissatisfied customers associated with a potential contractor. With the correct company name, the name of the owner, and the contractor's license number, your state contractor's license board, registrar of contractors, or bureau of consumer affairs should be able to supply you with information on adjucated cases against contractors or the complaint records of contractors. You should be able to find out if a potential contractor's license is in good standing, is on probation, has been suspended, or has been revoked.

- Invite three or four companies you have researched to provide estimates for the work you want done. Legitimate contractors provide free estimates to homeowners.

- So that bids are accurate, be certain to have each salesperson bid on the same specifications, completion time, and materials. If three of four bids are similar in cost and the fourth is much lower, be very cautious of the low bid. "Low-ball" contractors often operate without insurance or skimp on materials, equipment, labor, or effort to keep prices down. On the other hand, new companies might bid low to attract customers.

- Avoid contractors who want more than a 10 percent partial payment before starting the work. Established companies require no up-front payments from homeowners. Smaller companies might need a limited advance to cover the cost of purchasing materials.

- Ask to see copies of the contractor's liability insurance, workers' compensation insurance (an ACORD 25-S certificate), or (for an individual moonlighting odd jobs) personal disability insurance. Some states exempt liability or workers' compensation insurance requirements for sole proprietors and contractors with fewer than three or five employees. Some insurance experts advise that homeowners contact the contractor's insurance company to confirm that a policy is valid. Most insurers will send a copy of an insurance certificate to a customer if the contractor makes the request. To find out if your state will insure a homeowner for workers' compensation, telephone (800) 942-4242.

- Ask to see state, county, and local roofing contractor licenses. Contractor licensing requirements vary considerably throughout the country. Only 25 states require home-improvement contractors to be licensed. To obtain a license, a contractor might be required to pass a business management test and perhaps a trade test. Many license application classifications require listing a specific length of experience as well as the contractor's work history. Contracting without a license could result in a misdemeanor or, with repeated violations, a felony.

- Ensure that you have written warranties on materials and workmanship and that the contractor is bonded. Generally, a *limited product warranty* is provided for by the manufacturer of the roofing materials as long as the materials are installed according to the manufacturer's specifications. Warranties should be listed in your contract. Some states have a *recovery fund* financed by licensed residential contractors. Such funds are designed to cover losses to homeowners who have damages caused by failure of a licensed contractor to adequately improve the property.

The American Homeowners Foundation offers a six-page model contract for $6.95 that will help you avoid the pitfalls common to home-improvement projects. One excellent feature of the model contract is a sample clause specifying arbitration rather than litigation for settlement of potential disputes. See the Resources section in this book for the address and telephone number to check the current price and availability for the model contract.

Roof-It-Yourself Safety Guidelines

If you decide to do the work yourself, keep the following safety points in mind:

- Wear comfortable work clothes and soft-soled shoes with excellent traction. Don't wear jewelry, a watch, or leather-soled shoes.
- Roofing your home could easily mean lifting several tons of roofing materials. Work at a pace that is compatible with your health and conditioning; don't overextend yourself.
- During warm weather, take frequent breaks and drink a lot of water.
- Stay off the roof during very hot or very cold weather, when it rains, or when it is very windy. Don't attempt to walk on a roof that has frost, ice, or snow on it. Frost is especially dangerous because it isn't always easy to see on a roof.
- Use a very sturdy ladder placed at the proper angle and extended over the eaves by 3 feet. Keep ladders away from power lines. Refer to the section on ladders in chapter 2 for additional information on ladder selection and safe use.
- If you drop something while you are working on the roof, don't attempt to retrieve it. One more trip down and up the ladder is better than a headlong trip off the roof.

- Don't overload one section of the roof with materials.

- Never walk on felt that has not been securely nailed or stapled to the roof deck.

- Never leave scraps and shingle wrappers scattered over the roof. Toss trash in one pile on the ground.

- Unless you are already on the ground, don't step back to admire your work.

Estimating Materials

The first step in estimating materials for your roof is to determine the total number of square feet of the roof. The simplest way to find the total area is to climb onto the roof and take careful measurements. Be sure to wear shoes with soft soles, use safe ladder procedures, and take a tape measure, a pencil, and paper with you.

The design of your house is a key factor in how simple or difficult it will be to accurately estimate the amount of materials you will need. Simple rectangular shapes without dormers, valleys, or other obstacles and complications are computed by first measuring the *rake* (the parallel edges) of the building, then across the *eaves*, and then multiplying the two figures. The result is the total square feet of half the roof. Double the number for the approximate total square feet of the roof. The following figures can be used as an example for computing the total square feet of a simple rectangular roof surface:

rake:	16	feet
eaves:	× 50	feet
subtotal:	800	feet
	× 2	
total:	1600	square feet
	(16 squares)	

When valleys, dormers, and hip roofs are part of the structure, measuring gets a bit more complicated. For more complicated roof structures, partitioning large areas of the roof into squares or rectangles makes measuring and estimating easier. Remember that you are making an estimate. You should not try to figure down to the last shingle, the last piece of flashing, and the very last nail. A reasonable guideline is to allow an additional 10 percent for waste material.

In addition to finding the total square feet of the roof surface, you'll have to calculate the amount of material needed for *capping shingles* and *border shingles*. To determine the number of shingles needed for *ridge capping*, divide the total one-way length of the eaves measurement (plus the length of any hips) by 3, then multiply that total by 7. Divide by 3 because a shingle is 3 feet long and multiply by 7 because it takes seven individual caps to cover the length of one installed shingle. Note that each three-tab shingle will produce three cut caps. For more information on how to prepare capping and install border shingles, see chapter 5.

The following is an example of how to compute capping and border-shingle requirements.

eaves measurement:	50 feet
hip measurement:	+0 feet
subtotal:	50 feet
after dividing by three:	17 (round up)
	×7
total number of capping shingles:	119 (about 4$^1/_3$ bundles)

Border shingles are applied to the outer edges of the roof to provide additional protection (essential on the bottom course) and to present an even edge, as seen from the ground, at the rakes. To calculate the number of border shingles needed, add each of the rake measurements, plus twice the eaves measurement, and divide that total by 3. Here is an example:

eaves measurement:	50 feet
	× 2
subtotal:	100 feet
	16 feet
	× 4
subtotal:	72 feet
total:	172 feet
dividing by 3:	57 border shingles

The next step is to add the number of capping shingles to the number of border shingles. The total in this example is 176 shingles. This total is divided by 27 because that is the number of shingles in a typical bundle. Note that the number of shingles in a bundle varies with the type and weight of shingles or other roofing materials you purchase. Use the following as an example:

capping shingles:	119
border shingles:	+57
total:	176
divided by 27:	6$^1/_2$ bundles (233 square feet; 3 bundles equal 1 square)

To find the total number of squares for the estimate, add the total square feet of both sides of the building (1600 square feet), the total square feet needed for capping and border shingles (233 square feet), and 10 percent for waste (183 square feet). The total of 2016 square feet means that you should order 20$^1/_3$ squares of shingles from your supplier.

Remember to add the cost of nails, roofing cement, valley flashing, felt, and tools to the estimated total cost. Use the following descriptions of materials as guidelines.

Nails

Use hot-dipped galvanized, hot-galvanized roofing nails, or aluminum roofing nails. Roofing nails should be 11- or 12-gauge nails with $3/8$-inch heads.
To apply three-tab shingles using four nails per shingle, you will need
approximately 2 pounds of nails per square of shingles. Nails should penetrate $3/4$ of an inch into the deck of the roof. See Table 3-1 for recommended nail lengths to use on roof surfaces.

Table 3-1 Recommended Nail Lengths.

Nail lengths	*Roofing materials*
1"	Asphalt or fiberglass shingles on a new deck
1"	Roll roofing on a new deck
$1^1/_4$"	Reroofing over old asphalt shingles
$1^3/_4$" or 2"	Reroofing over wood shingles

Nails have inch, penny, and gauge designations. The abbreviation d
stands for the penny sizing of some nails. The term *penny* originated in
England where handmade nails were sold for pennies per hundred. For
example, 100 1-inch nails sold for two pennies; therefore, they became
known as 2d nails. The *gauge* of nails relates to the diameter of the wire
used to manufacture the nails. Roofing nails are 10 gauge.

Generally, roofing nails are sold in 1-pound, 5-pound, and 50-pound
boxes. There are approximately 600 aluminum roofing nails per pound, but
only about 200—because of their greater weight—galvanized steel roofing
nails per pound. You will need a little more than 200 nails per 100 square
feet of roofing area.

Galvanized roofing nails are mechanically plated, electroplated, hot-
dipped, and hot-galvanized. All four types of roofing nails start as standard
steel nails that are coated with zinc and bonded by a variety of techniques.
For mechanically plated nails, a thin coat of zinc dust—and sometimes an
additional very thin coat of chromate—is added. Electroplated nails—commonly used in pneumatic nail guns—are created using an electrolytic solution to provide a thin coating. *Hot-dipped galvanized nails* are placed
once or dipped a second time in molten zinc that is hot enough to form an
alloy on the outer layer of the steel. *Hot-galvanized nails* are made by placing small chips of zinc in a drum that is heated and rotated. As the drum
tumbles, the zinc chips melt and coat (sometimes unevenly) the nails.

Except for perhaps on homes located immediately next to saltwater,
galvanized nails can be used on roofs where standard roof installation techniques—each course of roofing material is blind nailed—are used. If ocean
spray or mist might be a potential problem, you can use type 316 molybde-
num stainless steel nails.

Roofing Cement

A variety of asphalt coatings, adhesives, and cements is available for roofing. Asphalt/plastic-based cement is the most frequently used product for waterproofing around chimneys, vents, and skylights. A gallon or two of roofing cement should be more than adequate for the typical 20-square home. Use roofing cement that specifies a plastic base or a rubberized compound. As an alternative to messy-to-use roof cement, the 3M company has developed a sealer that can be pressed in place. This caulklike product comes in 15-foot, $7/_{16}$-inch-wide solid tape and is guaranteed for 20 years against cracking and peeling.

Valley and Eaves Flashing

Valleys can be protected with galvanized metal or with a mineral-surface product such as E-Z Roof (made by Tarmac Roofing Systems) roll roofing or a rubberized polyethylene, self-adhering membrane.

   To estimate the amount of metal valley material needed, measure the length of the valley along the roof's surface. If you cannot obtain the metal in one continuous roll, add 6 inches for each overlap and add another 8 inches for overlap at the ridge and eaves. Use the same procedure for estimating mineral-surface roll roofing. If you use roll roofing for valley flashing, double the amount of the measured length. Two layers of roll roofing are needed for adequate protection against leaks in valleys. Metal and mineral-surface roll roofing used for valleys should be at least 36 inches wide.

Step Flashing

If you cannot obtain precut aluminum *step flashing*, cut 7-×-10-inch pieces and bend each piece in half to make them 7 × 5 inches. Allow for a 2-inch overlap of each piece when you are measuring for step flashing that is to be installed along a wall or chimney.

Pipe Collars

For a reroofing job, carefully remove and reuse all *pipe collars*. If the collars and shingles were properly installed, the old shingles will have protected the collars from wear. For new work, measure the circumferences of the pipes and purchase either metal or polyurethane collars.

Felt

For new construction or where worn shingles have been torn off your roof, install No. 15 asphalt-saturated *felt*, or *building paper*, as an underlayment for the new asphalt-fiberglass shingles. Each roll of felt covers 400 square feet. Some suppliers also stock No. 30 or No. 45 saturated felt, but the added thickness and expense are not required unless you are installing clay tiles or cement tiles.

   Felt is a necessary vapor barrier between the wood deck and the shingles. Otherwise, during hot weather, the wood deck warmed by trapped attic air will draw moisture from the shingles and prematurely age the shingles. If you will not be removing worn shingles, there definitely is no need to install a layer of felt when you are applying a second layer of shingles over the top of one layer of worn asphalt shingles.

Drip Edge

For new construction or when worn shingles have been removed, install medium-weight metal drip edge along the eaves and rakes. Total the eaves and rake measurements and add 10 percent for overlapping and waste. Drip edge is sold in 10-foot lengths.

Tools

You will need a roofer's hatchet (or a carpenter's hammer for non-purists), a nail pouch or nail apron, a chalk line and a tube of chalk, a utility knife and hook blades (or straight blades if you can't obtain hook blades), a tape measure, a trowel, tin snips, rubber-soled shoes, and a sturdy ladder. If you will be tearing off the old shingles, a flat-edged, tear-off shovel is essential. Do not try to tear off shingles with a garden shovel. If the pitch of the roof is steep, you will also need roof jacks and planking.

Consider using electric-powered nailers or pneumatic-powered staplers and a compressor if such equipment is available for rent. Most inexperienced roofers are lucky to be able to nail one square an hour by hand. A powered staple gun will double your shingling speed.

Nail Selection and Use

To install three-tab, asphalt fiberglass-based shingles on a new deck, use 1-inch galvanized or aluminum roofing nails (see Tables 3-2 and 3-3). Use $1^1/_4$-inch galvanized or aluminum roofing nails to install three-tab shingles over one layer of worn asphalt shingles. If you plan to install shingles over roof-deck insulating building materials, such as Homasote or Thermasote, you will need $1^3/_4$-inch to $2^1/_2$-inch roofing nails. In order for the nails to hold properly, the nails you use should penetrate twice the thickness of the combined insulation, sheathing, and roofing materials.

Longer nails are difficult to drive and require more time and effort to install than 1- or $1^1/_4$-inch roofing nails. An experienced roofer can drive 1-inch nails with one stroke of a roofing hatchet and $1^1/_4$-inch nails with one, or at most, two strokes. Longer nails require several strokes each. Such extra efforts might not seem significant until you consider how many nails it takes to roof a building.

➡ **Application Tip** Almost everyone takes for granted that they know the proper way to hold and drive nails. Most people use the technique favored by carpenters. The nail is held near the nail head between your thumb and forefinger. This method is fine for carpenters but for roofers it

Table 3-2 Galvanized Roofing Nails.

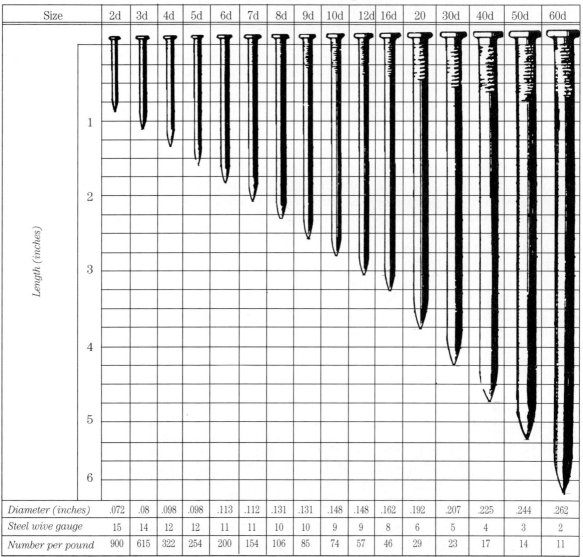

Size	2d	3d	4d	5d	6d	7d	8d	9d	10d	12d	16d	20	30d	40d	50d	60d
Diameter (inches)	.072	.08	.098	.098	.113	.112	.131	.131	.148	.148	.162	.192	.207	.225	.244	.262
Steel wive gauge	15	14	12	12	11	11	10	10	9	9	8	6	5	4	3	2
Number per pound	900	615	322	254	200	154	106	85	74	57	46	29	23	17	14	11

Table 3-3 Aluminum Roofing Nails.

Length	Gauge	Head diameter	Nails per pound	Nails per box	Shingle coverage per box
1"	10	$^{7}/_{16}$"	605	980	300 square feet
1$^{1}/_{4}$"	10	$^{7}/_{16}$"	491	980	300 square feet
1$^{1}/_{2}$"	10	$^{7}/_{16}$"	417	980	300 square feet
1$^{3}/_{4}$"	10	$^{7}/_{16}$"	368	650	200 square feet
2"	10	$^{7}/_{16}$"	336	650	200 square feet
2$^{1}/_{2}$"	10	$^{7}/_{16}$"	274	325	100 square feet

is too slow. It is also an easy way to pinch your fingers. In addition, it is very difficult to hold more than one nail in your hand when you use this technique.

Nailing shingles with a roofing hatchet is time-consuming and repetitious work. Each full shingle must have four nails. Each time, try to take four to six nails from your nail pouch and drive them before returning your hand to the pouch for more nails.

Hold the nails in the palm of your hand. Cup your hand and use your thumb to "roll" the nails between your fingers. Lay your knuckles against the deck and practice nailing. The heads of the nails will be easy to tap with the hatchet and you don't have to worry about pinching your fingers. It is not important which fingers the nails are rolled between, but it is important that the points of the nails be at the proper angle for setting.

The secret of the technique for rolling nails between your fingers is to position the nail properly and tap the nail head very lightly with your hatchet head so that the nail just remains upright in the shingle. Take your hand away from the nail and drive the nail with one stroke of the hatchet.

The nail head should be flush with the surface of the shingle. Do not drive the nail head deep into the shingle and do not leave it less than flush with the shingle. The idea is to develop a rhythmic motion that allows you to set a nail by tapping it, move your hand, drive the nail, and go to the next nail.

It is crucial that you install nails or staples at the proper locations (Figs. 3-1 through 3-3) for every shingle. Each shingle has a series of seal-

Fig. 3-1   When nails are properly located, the shingle courses will be double-nailed.

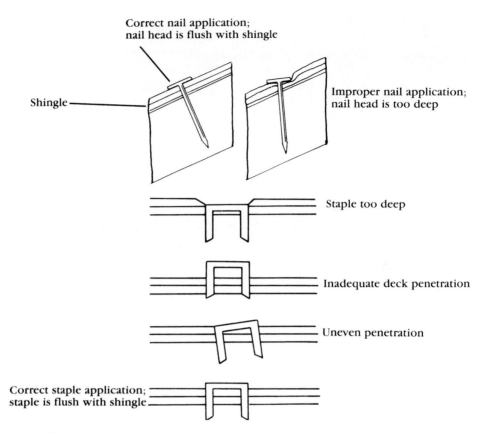

Correct nail application;
nail head is flush with shingle

Shingle

Improper nail application;
nail head is too deep

Staple too deep

Inadequate deck penetration

Uneven penetration

Correct staple application;
staple is flush with shingle

Fig. 3-2 Nails or staples must be driven flush with the shingle surface at the proper locations.

down adhesive strips on the face of the shingle. Nails must be driven below these strips, but above the *slots* or *cutouts* that make up the *tabs* of the shingle. If you drive nails above the adhesive strips, the nails will miss the shingle course underneath and you will not have double-nailed the courses. It is essential to have the double-nailed installation of the shingle courses. On the other hand, nailing too low—below the *watermark* that defines the amount of shingle exposed to the weather—will expose the nail heads and leaks will occur.

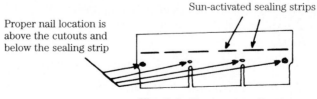

Sun-activated sealing strips

Proper nail location is
above the cutouts and
below the sealing strip

Fig. 3-3 Fastener application.

Fig. 3-4 If the shingle supplier will deliver the material to the roof, ask the driver of the truck for advice on where to position the boom.

Shingle Delivery

Getting the bundles of new shingles on your roof requires considerable effort. If your health and physical condition are less than excellent, you should not attempt such strenuous work. Although you will be lifting several tons of material, there are several pieces of equipment and several techniques you can use to make the labor easier to accomplish.

When you call local roofing material suppliers, you will probably discover that some companies:

- Have the type of shingle you want, but do not have delivery trucks.

- Will deliver shingles to a job site, but do not have the equipment to place them on rooftops.

• Will deliver shingles and have the equipment, such as a laddeveyor or conveyor belt, to reach the rooftops of one- or two-story structures but might charge extra.

The added cost for delivering shingles to the rooftop is definitely worthwhile. In most such cases, the driver of the delivery truck will place the bundles on a conveyor belt, and you will have to stack the bundles on the roof (Figs. 3-4 and 3- 5). Some companies will load the shingles on your roof as part of the material-delivery service charge.

Bundles of shingles are often warehoused on pallets stacked under thousands of pounds of pressure from their own weight. On rare occasions, a supplier will stack warehoused pallets of shingles under more weight than is specified by the shingle manufacturer. As a result, the individual shingles within the bundles compress, the protective tapes become ineffective, and

Fig. 3-5 The shingles are stacked at the eaves because the carpenters were not yet finished sheathing the roof.

tar and oils from the shingles bond or stick together and become a damaged or unusable mass.

A brief inspection of the materials while they are still on the delivery truck could save you considerable delays and time and effort insisting that the supplier replace materials that have been delivered to your rooftop. Ask the delivery driver for permission to open a bundle so that you can determine if the shingles will come out freely. While you are at it, measure a shingle to confirm that the dimensions are exactly the same as specified on the bundle wrapper. Check that the materials about to be delivered exactly match the brand, style, weight, and color (beware of mixed blend numbers) you ordered from the supplier.

When shingles are delivered to the job site but not to the roof of the building, the ideal storage method is to stack the bundles not more than 4 feet high on wooden pallets. If you will not be immediately carrying the bundles onto the roof, cover the shingles with a tarpaulin. Shingles should not be saturated by rainwater immediately before application. Keeping the paper wrappings dry makes carrying the bundles less difficult than if they are wet. The wrappings on wet bundles of shingles will split when you try to lift the bundles. Also, carrying dripping-wet shingles up a ladder makes a difficult task a very unpleasant one.

You should not attempt to carry dozens of bundles of shingles unless you are free of back troubles, have excellent health, and are in excellent condition. When you lift a bundle from a stack on the ground, guide it with one arm so that the shingles will land flat on your shoulder, and let momentum and gravity topple the bundle from the stack to your shoulder.

The bundle should be centered on your shoulder. Use your free arm to grasp the side of the ladder as you climb the rungs until your shoulders are about 2 feet above the eaves. Gently lower the bundle to the roof. By making two more such trips, you will have plenty of material to start work. Carrying too many bundles to the roof at one time is exhausting labor and places too many bundles on only one area of the roof.

As you reduce the number of bundles stacked on the ground, it will become more difficult to get the bundles from the stack to your shoulder. An efficient technique for lifting the bundles is to:

- Lean the bundle against your thigh.
- Grasp the sides of the bundle with your hands.
- Bend your knees slightly.
- Pull with your arms.
- Flip up (not end over end) the bundle to your shoulder. This technique takes some practice before it becomes routine. Don't try it with wet or broken bundles.

If the roof you will be working on is not steep and your roofing-material supplier has the equipment for delivering the shingles to the roof, you

Fig. 3-6 At the ridge line, position the first two bundles to form a platform.

should stack the bundles across the *peak* or *ridge* of the building (Figs. 3-6 and 3-7). Ask the driver of the delivery truck for advice on positioning the conveyor belt. It will probably be necessary to position several bundles or planks under the top of the conveyor-belt mechanism as it rests on the roof surface so that the belt travels freely (Fig. 3-8). It will also make it easier to catch the bundles.

After all the equipment has been set up and the first bundle has been caught, walk to the ridge of the roof and position this bundle so that it is approximately 6 inches below the ridge (Fig. 3-6). The bundle must be placed parallel with the peak. Position the next bundle so that it creates a platform for four or five more bundles. Repeat this pattern until you have stacked all the shingles on the roof. This technique will allow you to stack the bundles flat; they will not slide off the roof.

Fig. 3-7 Stacking the bundles at the ridge line provides room to work, and the shingles won't slide.

Fig. 3-8 Raising the boom off the deck allows room for the conveyor mechanism to operate and makes lifting the bundles easier.

If your roof is too steep for stacking bundles at the ridge line, roof jacks and planking can be installed to hold several bundles at a time. Do not overload the jacks and planking. Lay the shingles against the roof and use the planking to prevent the bundles from sliding (see Figs. 2-13 and 5-49).

Chapter **4**

Tearing Off and Drying In

*W*orn shingles (Fig. 4-1) should be torn off if the shingles are warped and curled (Fig. 4-2) and the new roofing cannot be applied evenly over the old shingles. If your home already has two layers of worn shingles—even if the local or national building codes indicates otherwise—tear off all the old layers of shingles and felt and make any repairs to the roof deck before you install new shingles over new felt. Manufacturers will not guarantee their products if shingles are used to reroof over more than one layer of worn shingles.

Don't believe it if someone tells you that you can get away with applying three or more layers of shingles on your roof. Even if the top layer of shingles is not badly curled or warped, standard roofing nails will not adequately penetrate the roof deck through more than two layers of shingles. Some local building codes prohibit the application of more than two layers of shingles because of the difficulty of chopping through several layers of shingles if there is a fire. The extra weight of a third layer of shingles is another reason why no more than two shingle layers should be applied on a home. Many residential buildings are not designed to handle the excessive stress caused by what would be the addition of several more tons of roofing material.

Be prepared to cover items stored in your attic. During tear-off work, grit, dirt, dust, nails, and other debris will fall through the gaps and knotholes in roof-deck planking and sheathing.

Tearing off one or two layers of asphalt roofing shingles, wood shingles or shakes (Figs. 4-3 through 4-5), gravel, felt, and tar (Figs. 4-6 through 4-8), or roll roofing (Fig. 4-9) are difficult and physically demanding jobs. While it is possible for one person to do all the work, it will be more realistic, practical, safer, and more economical for you to arrange to have several helpers assist with your project.

Fig. 4-1   Worn shingles often show a significant loss of color granules.

Have a sturdy truck or construction-debris disposal container available to haul away several tons of material. If possible, park the truck or locate the container under the eaves and push the worn material directly from the roof onto the truck bed (Figs. 4-8, 4-10, and 4-11) or into the container (Fig. 4-12). Be sure to research local landfill restrictions and check local government and business agencies to see if your community has one of the few businesses that will recycle some of the more than 1 billion squares of roofing materials torn from roofs each year. See the Resources list in this book.

Don't attempt to take on more work than you can reasonably handle in one day. Plan to tear off the shingles, dry in sections of the roof deck with felt, lay chalk lines, install drip edge, and reshingle only as much of the roof area as you can manage in one day's work.

Keep in mind the weather forecast for your area. If you give in to the temptation to keep working until you remove all of the old roofing material, a rainstorm just might drench everything in the middle of the night. If

Fig. 4-2 Curled or warped shingles aged by weather must be removed.

for some reason you find that you have uncovered more of the roof deck than you can reshingle the same day, carefully nail felt to *dry in* all sections of the exposed deck. Make sure there are no wrinkles or tears in the felt. Nail wood *furring strips* parallel with the layers of overlapped felt. Carefully secured felt will provide adequate waterproofing for days or even weeks in some climates. If your plans require the roof deck to be exposed for more than a few days, consider renting or purchasing enough tarpaulin to cover the roof.

Wear gloves when you are tearing off shingles, to prevent blisters. Begin tearing off shingles by slipping the edge of the shovel under the edge of a few pieces of capping (Figs. 4-13 and 4-14). After you have removed capping along the length of two or three shingles (Figs. 4-15 and 4-16), turn and face down the slope. Slip the edge of the shovel under the edge of the top of the first shingle course (Fig. 4-17). Pry the nails loose (Figs. 4-18 and 4-19) but don't try to pull each individual shingle out one at a time. Instead, work the shovel further down under the second course and pry more nails loose (Fig. 4-20).

Try to loosen as many nails as possible and "roll" several courses of shingles down the slope. Continually pry and push layers of shingles. Work

Fig. 4-3 A tear-off spading fork loosens layers of wood shingles.

Fig. 4-4 Claw hammers and pry bars remove nails and debris from the roof deck.

from the sides of the area you have cleared to get at the capping and top course of more shingles on one section of the roof. Always begin at the ridge and work down, letting gravity help you remove the debris (see Figs. 4-21 through 4-23).

Be sure to remove even the smallest pieces of shingles (Figs. 4-24 and 4-25). A pry bar (Fig. 2-27) will prove very useful. Pull all nails or drive the exposed nail heads into the deck (Figs. 4-4 and 4-26). Replace any damaged planking (Fig. 4-27 through 4-29, Figs. 14-8 through 14-13, and Figs. 14-26 through 14-36).

When you remove the shingles around a vent, work carefully and keep the *flange* for reuse. Galvanized or lead flanges almost always can be used again because the old shingles, if they were properly installed, will have protected the flanges from the weather over the years (Figs. 7-61 through 7-68). Reused or new flanges can be spray painted to match the color of the new shingles. See Figs. 12-44 through 12-55.

Fig. 4-5 Tons of debris results from tearing off wood shingles. Before you start work, make arrangements with your local landfill or recycler.

Fig. 4-6   Stone, tar, and felt can be torn off in large sections.

Fig. 4-7   Arrange for a crew of helpers for tear-off work.

Fig. 4-8 A scoop shovel helps place gravel directly into a truck bed.

Fig. 4-9 Layers of roll roofing and asphalt cement can be difficult to remove without spading forks.

Fig. 4-10 Using a wheelbarrow on a low-slope roof can reduce the number of trips to the debris-hauling truck.

Laying Felt

Use No. 15 saturated felt to provide a smooth surface on which to nail shingles and as temporary waterproofing in case of rain. A layer of felt must be applied to a wood deck to provide a *vapor barrier* between the wood deck and the asphalt-fiberglass shingles. Without the felt layer, the wood planking or plywood deck—dried by superheated temperatures found in most attics—will draw moisture (including tar and oils from the asphalt) from shingles. The result will be shingles that are prematurely worn, cracked, brittle, as well as stuck to the wood surface when you attempt to remove them.

Beginning at a bottom corner of the roof, lay felt courses horizontally. Carefully cut off the binding—but don't cut into the roll—and unravel 2 or 3 feet of the roll. Position the roll of felt so that it can be rolled across the bottom of the roof, even with the eaves. Kneel and hold the roll in both hands. Maneuver the felt into position so that it covers the deck right up to the edge of the rake and eaves but not over the sides of the building. See Figs. 4-30 and 4-31.

When you are satisfied that the roll is positioned properly, drive about five nails into the top right-hand corner of the felt. Roll out the felt no more than halfway (perhaps 25 feet or so) across the average-size roof. If there is any wind at all, roll out the felt only about 10 feet.

Fig. 4-11 Position a truck as close as you can to the eaves and aim the debris for the truck bed.

Never walk on felt that has not been nailed down. Pick up the roll with both hands, pull, straighten, and align the felt along the eaves. Be sure that there are no wrinkles. From behind the roll, reach over and nail down the top of the strip with roofing nails spaced 6 to 8 inches apart. Nail the middle and bottom of the felt similarly.

Roll out the felt toward the other end of the roof. Leave yourself enough room to pull the felt free of wrinkles and set it even with the eaves. Repeat the nailing pattern and remember not to walk on unnailed felt. Unravel a few more feet of felt and cut it with a utility knife. Trim any felt that overlaps the rake. Nail down the last few feet of the first course of felt.

Position the next course of felt so that there is a 2-inch overlap of the first course of felt (Figs. 4-32 and 4-33). Use the white lines printed on the felt as guidelines for lining up the courses. The bottom of the second course of felt should be on top of the first course so that any moisture will flow over the layers of felt.

Roll out a few feet of felt and align the edge of the felt along the rake and the first course (Fig. 4-34). Remember to leave a 2-inch overlap. Drive about five nails in the top right-hand corner and roll out the felt about halfway (perhaps 25 feet) across the roof. Stand on the first course of felt and work with the second course of felt above you. It is not necessary to completely nail the top edge of the course. A few nails to hold the top in

Fig. 4-12   Consider renting a portable construction debris container for hauling away worn shingles.

place will be adequate because each top row will become a bottom row once you add another course of felt.

Nail down the bottom and middle of the second course of felt with a pattern of nails every 6 to 8 inches (Fig. 4-35). Remember: never walk on felt that has not been nailed down. Continue laying felt over the remainder of the deck using the same techniques (Fig. 4-36). When you reach the top course, lap about 6 inches of felt over the ridge top (Fig. 4-37). The longer the deck will be exposed to the weather before shingles are applied, the more important it is to lay the felt so that the deck is watertight.

Using a utility knife, slice any wrinkles and nail the felt so that it is smooth. If the deck will be exposed to the weather overnight or longer, apply a very thin coat of asphalt-based roofing cement to waterproof the areas where cuts have been made.

➡ **Application Tip**   Do not use roofing cement to patch newly installed felt if you will be reshingling the deck the same day. Cement takes several hours to dry and wet cement might ruin your chalk line when you snap

Fig. 4-13 Begin tear-off work at the ridge.

Fig. 4-14 Slip the edge of the tear-off shovel under several pieces of capping.

Fig. 4-15   Pry away both layers of shingles and several pieces of capping.

Fig. 4-16   Remove two or three shingles.

Fig. 4-17 Loosen the edge of the top layer.

Fig. 4-18 Pry away at the second layer of shingles.

Fig. 4-19   Find the nail heads with the flat edge of the shovel and pry.

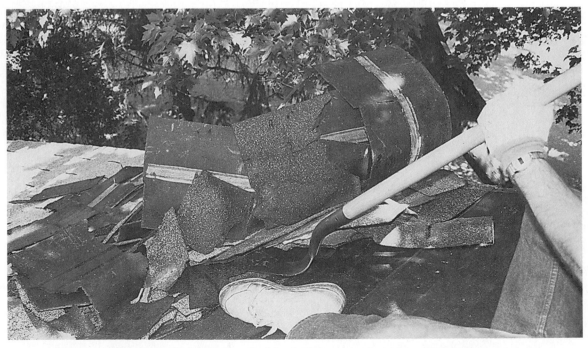

Fig. 4-20   Work the shovel under both layers and pry more nails loose.

Fig. 4-21 Clear the first few courses across the ridge.

Fig. 4-22 Work down the roof section by tearing large swaths of both layers of worn shingles.

Fig. 4-23 Allow gravity to help you remove debris.

Fig. 4-24 Sweep the deck to remove even the smallest pieces of worn shingles and nails.

Fig. 4-25 A bucket is useful for collecting the many small pieces of worn shingles that are torn from a roof.

lines over the areas. Also, freshly laid asphalt-based cement would certainly stick to the line and make it impossible to rewind. If there are cemented areas that must have chalk lines over them, nail a piece of scrap felt over the cemented areas. The deck will be waterproof and the chalk line will remain clean.

When you lay felt in a valley, make certain there are no rips or large wrinkles in the felt. Before you apply the horizontal sections of felt, install a vertical length of felt down the center of the valley. Several sections of felt can be used, but be certain to overlap the higher sections several inches so that any water will run over the top of the felt.

Very carefully cut off the horizontal sections of felt at an angle as you reach the center of the valley. If the roof will be exposed overnight or longer, cover nail heads in the valley with a dab of roofing cement.

Installing Drip Edge

On homes where the eaves and rakes are visible from the ground, *drip edge* is an attractive way to set off the area where siding and shingles meet. Medium-gauge, aluminum drip edge is inexpensive and easy to install. Don't use heavy-gauge, galvanized-metal drip edge that is designed for use on hot-tar-and-stone, build-up roofs or drip edge that is so lightweight that it is extremely flimsy. The chief function of drip edge is cosmetic, but it

Fig. 4-26 Watch closely for any remaining nails that must be pulled or driven into the deck.

Fig. 4-27 Wood shingles were torn from the roof of this building. New plywood, felt, and fiberglass-based shingles were installed. The height of the building and the pitch of the roof make this a job for professional carpenters and roofers.

Fig. 4-28 Planking can be removed temporarily to allow access to worn flashing and roofing.

Fig. 4-29 Damaged planking is replaced with new boards of equal thickness attached to the nearest rafter or truss.

Fig. 4-30 Position the roll of felt and drive three nails at the top right corner.

Fig. 4-31 Roll out the felt, align it along the eaves, pull it taught, and drive one nail at the top right corner. Continue across the roof section.

Fig. 4-32 With a new roll of felt, trim the folded portion before you install the next course.

Fig. 4-33 Overlap the courses of felt by 2 inches with the second layer on top.

Fig. 4-34 Roll out the second course of felt and drive a pattern of roof nails.

does provide a few inches of additional protection against wind-driven moisture and potential pests such as carpenter bees where the deck and eaves meet.

Drip edge is installed with the 3-inch flat portion against the deck and the creased portion curled snug against the edge of the deck. The 10-foot sections of drip edge are fastened with roofing nails driven into the deck about every 4 feet. Where you overlap sections of drip edge, make sure that the length of drip edge closest to the ridge of the roof is on top in order to ensure the most attractive appearance when viewed from the ground.

To obtain the best results when you need to cut the drip edge, always first trim the rake edge (Fig. 4-38), set the drip edge in place—without nailing it down—and then use a nail to scratch a line where a cut is needed. Then cut the drip edge, position it, and nail it down.

Where the rake and eaves meet, use tin snips to cut (from both ends) only the flat portion (Figs. 4-39 and 4-40) of the drip edge so that it will wrap around a corner (Fig. 4-41 and 4-42). At the ridge, be sure to make one cut only on the outside (curled) portion of the drip edge (Figs. 4-43 and 4-44) to eliminate the bulge that would otherwise remain. Drape the remainder of the section over the top of the ridge and down the other rake. Be sure to lap the flat section of the drip edge over the top of the next full-length piece of drip edge (Figs. 4-45 and 4-46).

Fig. 4-35 Where sections of the felt overlap, drive nails every 3 or 4 inches. Note how the nail is held with the palm of the hand up.

From the ground, look for any loose-fitting or improperly installed drip edge. Making corrections at this point is considerably easier than after you have shingled the entire roof.

➡ **Application Tip** Many roofers install strips of drip edge on top of the starter course of felt at the eaves (Fig. 5-31) to keep the felt secure from wind gusts. Others insist that drip edge installed along the eaves must be placed under the starter course of felt so that any moisture that theoretically is able to penetrate overlapping courses of shingles will not be trapped by the lip of the 3-inch-wide drip edge at the eaves.

Those who feel that the eaves' drip edge must be installed under the first course of felt seem to ignore the obvious. Either entire sections of the

Fig. 4-36 Continue installing felt courses until you reach the ridge.

Fig. 4-37 At the ridge, lap the felt over both sides by at least 8 inches.

Fig. 4-38 Before you install drip edge, use a utility knife to trim the felt at the rake.

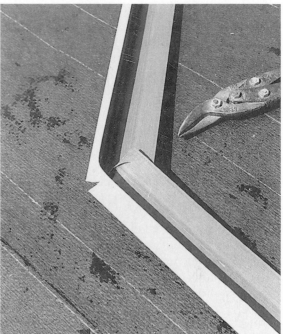

Fig. 4-39 Cut only the flat portions of the drip edge to install drip edge at the corner of a rake and eaves.

Fig. 4-40 To fit a corner, cut both flat portions of the drip edge.

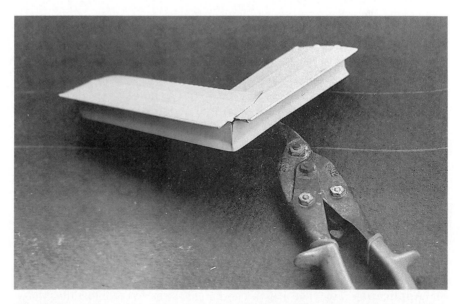

Fig. 4-41 At the corner, bend the bottom section of drip edge under the section that will be installed along the rake.

roof would have to fail or—somehow—only the first course and the eaves border shingles would have to fail in order for moisture to reach the bottom of the first course of felt and the 3-inch-wide drip edge at the eaves. Then, to cause a problem, the moisture would have to be trapped rather than flow over the drip edge. If your drip edge is trapping enough rainwater runoff to cause any damage to your roof deck, your roof wore itself out a very long time ago.

Everyone seems to agree that drip edge is correctly applied on top of the felt at the rakes (Fig. 5-32).

Laying Chalk Lines for Three-Tab Shingles

Vertical and horizontal chalk lines are needed for the proper application of three-tab (36-inch-long) shingles where the cutouts are aligned every other course as you look toward the peak of the roof. If you purchase shingles manufactured to metric dimensions, you will have to adjust your chalk line measurements to meet the requirements listed on the shingle packages and you might have to follow specific application instructions provided by the manufacturer. The best way to apply metric shingles is to use the random-spacing pattern that essentially eliminates any concerns about the alignment of cutouts.

At the corners and the center of the first section of roof you are preparing to shingle, begin laying chalk lines by measuring for the bottom course of border shingles. Use the following procedures for a uniform, 1-inch overhang along the eaves and rakes of your home.

Extend 1 inch of the tape measure over the edge of the drip edge (or the edge of the eaves if drip edge is not used) and scratch a V in the felt exactly at the 12-inch mark. At both corners, make sure that the points of the V marks are scratched within 6 inches of the edges of the rakes.

Where you make the 12-inch mark along the eaves near the center of the roof section, drive a roofing nail about a $1/2$-inch deep into the deck at the point of the V. Remember that 1 inch overlaps the drip edge when you confirm that the 12-inch nail point is correctly positioned.

Place the loop that is on the end of the chalk line over the nail head (Fig. 4-47) and walk toward either the right or the left corner of the roof. Slowly unwind the chalk line, and don't let it drop or scrape against the deck. Wrap the line around your index finger and pull the line taut. Hold the line against the point of the mark and very lightly snap it once (Fig. 4-48). Rewind the line and walk toward the other corner. Pull out enough line and repeat the snapping procedure.

➡ **Application Tip** The 1-inch overhang is intended to allow meltwater and rainwater runoff to flow into the eaves' gutters and to provide an attractive, even appearance at the rakes as seen from the ground. While

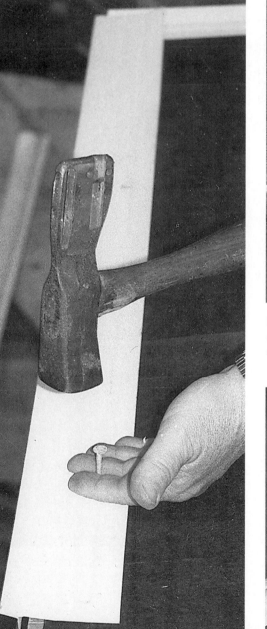

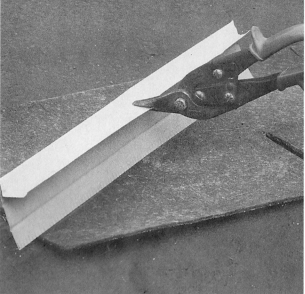

Fig. 4-43 After you position the drip edge at the ridge and carefully determine where to cut, use tin snips to cut only the outside, curled portion of the drip edge.

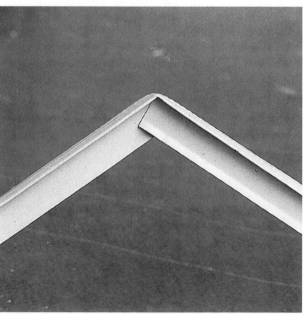

Fig. 4-42 Install drip edge by driving nails about every 2 to 4 feet along the rake.

Fig. 4-44 Bend the drip edge for a neat appearance at the ridge, and nail the metal in place.

Fig. 4-45 The top piece of drip edge is lapped over the bottom section.

Fig. 4-46 Carefully align the drip edge before you nail it in place.

some roofers install shingles cut even with the rakes or with less than a 1-inch overhang, roofs with a 1-inch overhang are far more common. Border shingles on eaves installed with less than a 1-inch drip edge overhang ignore the purpose of the drip edge and could contribute to leaks—due to *capillary action*—across the eaves. On a low-slope roof surface, the attraction or repulsion of water is due to the inherent surface tension of the solid roof surface and the liquid. The resulting backflow of rainwater does

Fig. 4-47 If you are working alone, you can snap chalk lines by looping one end of the line over the heads of nails driven part way into the deck at measured intervals.

Fig. 4-48 Hold the line against the deck and snap the line once.

not easily run off the last few courses of shingles and into the gutters, and the backflow can cause leaks.

Horizontal Lines

Measure the width of the shingles you are using. If they are $12^1/_8$ inches wide, measure and mark a V (start from the 12-inch mark where you have just snapped a line for the border shingle) every $5^1/_4$ inches for the first four lines and every $10^1/_4$ inches for additional horizontal lines the remainder of the way up the roof (Fig. 4-49).

If the shingles are 12 inches wide, mark a V every 5 inches, starting from the 12-inch chalk line for the first four lines and then at every 10 inches for additional horizontal lines. After the lines are laid, use a tape measure to check the accuracy of a few of the lines. Nailing shingles according to inaccurate lines could very easily throw off an otherwise fine-looking job. Lines properly laid will serve as excellent guides for keeping shingles straight.

Vertical Lines

As a reference point, the left and right corners in the following description are found by turning your back to the ground and looking toward the peak of the roof. Right-handed people should start at the lower left corner of the roof; left-handed people should start at the right corner. This way, you can keep the work in front of you as you nail the shingles.

Measure in from the rake (with the tape measure extended 1 inch over the drip edge) and scratch Vs at the 12-inch, 30-inch, and 36-inch marks at the first top corner of the first roof section you will shingle. Next, at the bottom corner of this first section, extend the tape measure 1 inch over the edge and scratch Vs at the 12-, 30-, and 36-inch marks. Now snap vertical chalk lines through these top and bottom marks and check the lines for accuracy.

It is extremely important that the 12-inch, 30-inch, and 36-inch vertical lines be accurate. Confirm the measurements for these lines at several places along the roof deck.

➡ **Application Tip** If you find a mistake after you have snapped your chalk lines, use a different chalk line with different-colored chalk to snap new lines. When one line is inaccurate it often affects the accuracy of other lines. As a result, using a different-colored chalk might be the only practical solution. If just the three vertical course-starter lines are inaccurate and you don't want to use different-colored chalk, take a nail and make a series of scratches in the felt every few inches along the length of the line, then snap new lines. The scratches help you to avoid getting the lines confused even though the lines are all the same color.

Fig. 4-49 Snap chalk lines to maintain an accurate application pattern.

Chapter **5**

Application Patterns

*T*hree common patterns are used for applying three-tab, asphalt fiberglass-based shingles. Shingle manufacturers usually specify that either the 45 pattern or the random-spacing pattern be used to ensure the best distribution of shingle color blend. The straight pattern for installing shingles is not prefered by product manaufacturers due to potential color variations in shingle lots. Once shingles are installed, on a roof, unattractive color variations resulting from mixed lots are sometimes noticeable—especially with white or pastel colors—in the stepped application patterns and are frequently visible, as the pattern moves from one bundle to the next, with the straight pattern. If your shingle order does contain mixed lots, avoid mixing lots on individual roof sections and use bundles with the most prevalent lot number (usually stamped on one end of each paperbound bundle) to cover the sections most visible from the street.

The straight pattern is often used to shingle dormers, catwalks, and other roof areas that are primarily vertically oriented surfaces. To shingle A-frame buildings and other roofs that are very steep, you must use roof jacks. When you use roof jacks, it is much easier to install shingles with the straight pattern because the work remains in front of you as you cover the roof and fewer movements of the planks and jacks are required.

Starting Points

If your roof consists only of squares and rectangles, selecting a starting point is easy. Working first on the back portion of the roof will give you the opportunity to become familiar with the application pattern and the time and effort needed to do the job. If the shingle cutouts are not perfectly aligned for your first attempt, it will not be as visible on the back of the house.

If you are right-handed, begin at the lower left corner of the back of the roof to keep the work in front of you. If you are left-handed, start work at the lower right corner of the back of the roof. The descriptions accompanying Figs. 5-1 through 5-25 are oriented for right-handed persons to simplify the instructions. Reverse the starting points—to the opposite side of the roof sections—if you are left-handed.

Fig. 5-1 Install shingles using the 45 pattern and begin at the front and back lower left-hand corners.

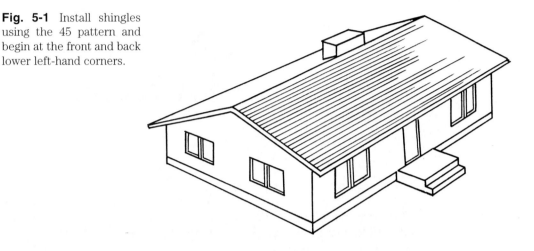

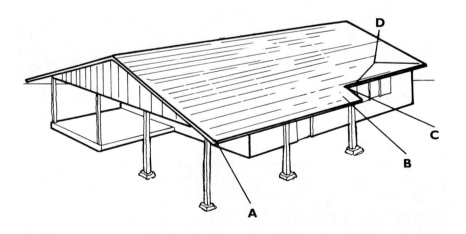

Fig. 5-2 Begin at the left-hand corner (A) and shingle to the angle (B). Be sure to install border shingles at the angle and continue the 45 pattern past the break in the roofline (C). If a short course is needed to maintain the proper exposure, make the adjustment at the course (D) that aligns with the break.

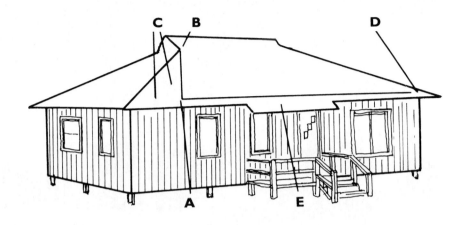

Fig. 5-3 Snap chalk lines between points A and B; fill in section C. Shingle across to point D but high-nail and back down across point E.

Fig. 5-4 Beginning at point A, use the 45 pattern to shingle past the angle, as shown in Fig. 5-2. On the second story, begin at point B. As you measure for chalk lines, take into account the chimney. If a chimney is at the ridge line (C)—where a left-hander would begin this section—measure for the vertical chalk lines just below the chimney and draw the lines through the marks so that they extend past the chimney to the ridge.

Fig. 5-5 Begin at point A, snap tie-in chalk lines at the dormers (B), and shingle the dormer tops by starting at the rakes (C).

Fig. 5-6 Starting at point A, shingle to the wall (B) and install step flashing. Start the second section at point C.

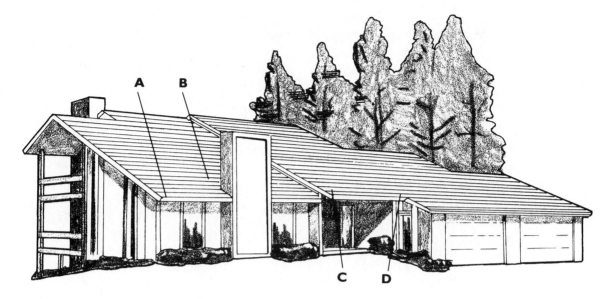

Fig. 5-7 At point A, begin shingling and use chalk lines to tie in the pattern around the chimney (B). Shingle across from point C and back down the pattern at point D by high-nailing the course (C).

Fig. 5-8 Start each of the three sections at the lower, left-hand corners. Make certain section C is not thrown off by a wrong measurement at the wall.

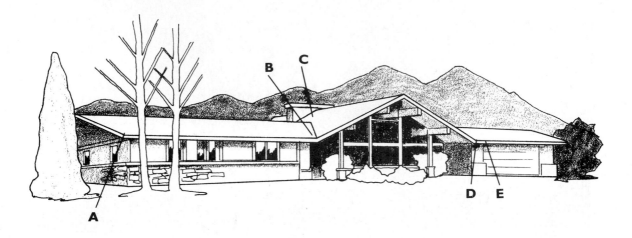

Fig. 5-9 Begin at point A and shingle toward the valley at point B. Work up the rake and toward the valley and chimney. At point C, back the shingle pattern down the roof. Start at the rake to shingle point D. At point E, start at the eaves, square off the section with chalk lines, and back in the pattern toward the valley.

Fig. 5-10 Begin at point A and tie in past the chimney. Shingle up the rake (B) and high-nail across to point C. Back down the bottom section of the main roof.

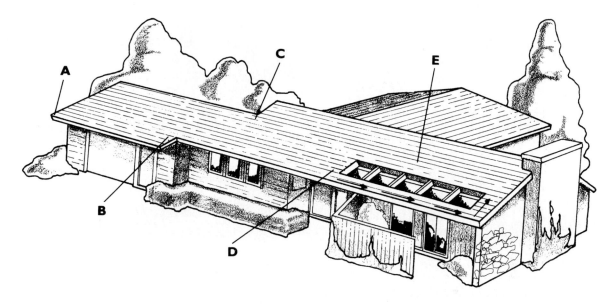

Fig. 5-11 Using the 45 pattern, begin at point A and shingle past the first angle. At point B, install border shingles along the rake and eaves, and lay a short course if necessary. At point C, snap chalk lines to continue the pattern. At point D, shingle past the bottom of the obstacle and up the rake. Snap chalk lines across point E to continue the shingle pattern.

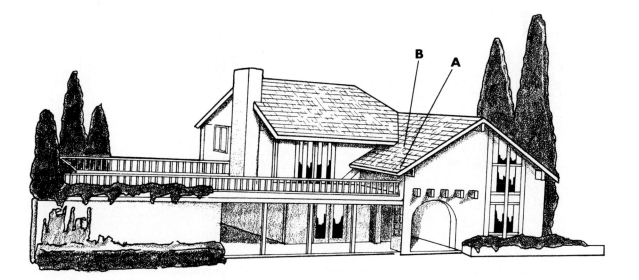

Fig. 5-12 Because of the valley and the wall, begin shingling at point A. Where the valley intersects (B), the valley material must go on top of at least one course of shingles.

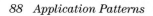

Fig. 5-13 Use roof jacks and planking on a roof that has a pitch this steep. Begin at the lower left corners of each section (A, C, D). Use roofing cement at the tabs to secure the shingles where the roof line steepens (B).

Fig. 5-14 Shingle the top section first (A). Square off the hip section (B) and shingle across the roof using the 45 pattern.

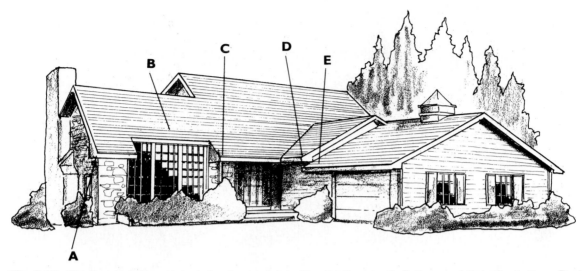

Fig. 5-15 Use the straight pattern to begin at point A. At point B, snap chalk lines and tie in the pattern. Back down the courses to the eaves (C) and shingle into the valley. Begin shingling at the rake at point D and at the eaves and wall at point E.

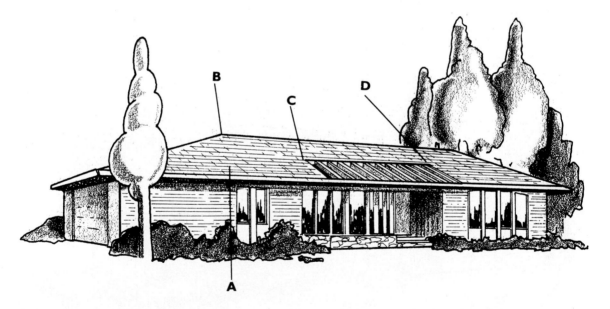

Fig. 5-16 Snap vertical chalk lines between points A and B. At point C, install a horizontal chalk line and border shingles across the top of the break in the roof line. Back down the shingle pattern at point D.

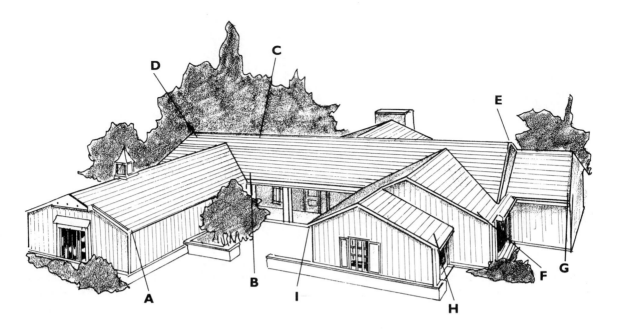

Fig. 5-17 Because of the valleys and rakes, this home design looks difficult to roof. Dividing the job into sections makes the work manageable. Start at point A and shingle toward the valley. Snap chalk lines between points B and C and fill in across to points D and E. Back down along the rakes and continue to fill in. Start at the rakes for the remaining sections (F through I).

Fig. 5-18 At first glance, it looks as if you should begin shingling this roof design at the middle of the house. However, starting at the middle would require a considerable amount of backing in. By starting at the left rake (A), you can keep the work in front of you. Be sure to install border shingles along the eaves and rakes where the roof lines are angled (B, C, and D).

Fig. 5-19 Shingle the entire top portion (A) of the building first so that you do not mar the shingles with a ladder extending from the lower section to the top section. Shingle the lower section starting at point B so that the pattern extends across and up the roofline (C).

Fig. 5-20 Although it means backing in, always work from the eaves toward the valley on this type of house design. Because the roof sections are not large, consider using the straight pattern (A, B). At point C, the bottom of the right side of the valley material must extend over the top of at least one course of shingles.

Fig. 5-21 Hip roofs require that you cut many shingles as you install the pattern. Where possible, square off the sections with chalk lines (A, B, C, and D).

Fig. 5-22 Begin work at the rakes (A, B). At the valley (C), install the bottom left corner of the valley material over the top of at least one course of shingles.

Fig. 5-23 Start shingling at point A and continue the border shingles and the pattern past the angled section (B).

Fig. 5-24 This home design would be easy to shingle if the pitch of the roof were not so steep. Roof jacks will be needed to do the work. Working from a ladder, begin at point A and shingle across the eaves. Install as many courses as possible from the ladder.

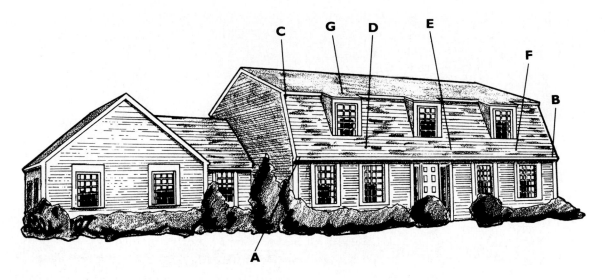

Fig. 5-25 Because mansard roof shingles are highly visible, consider having a professional roofer do the mansard work while you do the remainder of the roof. The mansard should be shingled, starting at point A, across the eaves (B). Continue the pattern up the rake (C) and the dormer walls. Snap vertical chalk lines (D, E, and F) to properly continue the pattern. On the top section of the roof, shingle across the eaves and back down the pattern at the dormer tops (G).

The starting-point suggestions shown in Figs. 5-1 through 5-25 will help you deal with intrusions in the roof. Depending on how you proceed, the sequence in which you shingle the sections of your roof will make the work easier or more difficult.

Shortcuts and Back-Saving Tips

A few roofing techniques can save you from extra effort, movement, or aggravation during shingle installation. Often, the first few shingles you install will have to be nailed into position while you are crouched in a somewhat awkward position facing the ground (Fig. 5-26). An alternative method is to work from a ladder (Fig. 5-27 and 5-28).

After the first few courses are on, sit on the roof deck with the work in front of you. If you are right-handed, tuck your left foot underneath your body and balance yourself so that you are comfortable. The center of your weight should be on your left hip. This position will give your upper torso freedom of movement. You should be able to nail at least four shingles before you have to move up the roof. After a little practice, you will get the hang of positioning yourself so that you will be close to the work but not too close.

Each full shingle must be *blind nailed* with four nails or staples positioned so that the fasteners are concealed by the shingle above it. As you

Fig. 5-26 Face the ground while nailing the first few courses at the rake.

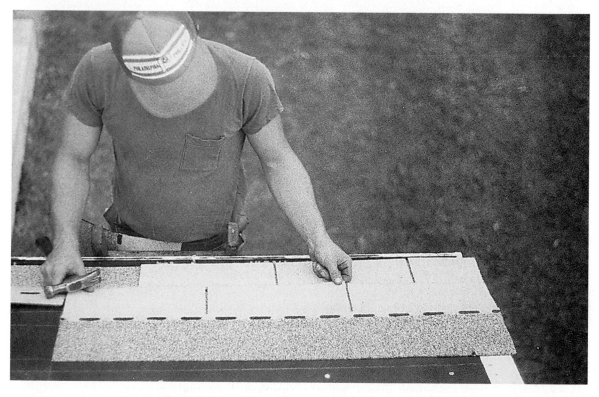

Fig. 5-27 Install the first few courses from a ladder.

Fig. 5-28 Don't attempt to reach farther than your arm's length when working from a ladder.

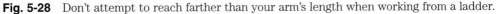

sit nailing shingles for each *course*, of the 45 pattern or the random-spacing pattern, you will soon discover that the outside nail on the pattern is the most difficult to reach. Instead of stretching for it, leave that outside tab temporarily unnailed and catch it on the next run. Each time you position a shingle, drive three nails into the shingle that you position and drive the fourth into the tab from the prior run at your left. Any nails that remain in your hand can be driven into the next few "fourth" tabs of the above courses.

Inexperienced roofers will sometimes contort into all sorts of uncomfortable stances when nailing shingles. Even some crews of professional shinglers—defying gravity and common sense—will work the entire roof from bottom to top with their backs to the peak and each course of shingles below their feet. Such positions will place considerable strain on your back. Don't try to kneel, bend over while standing flat-footed, or work the entire job "upside down."

Another aggravation to avoid is to needlessly pull off the plastic or adhesive protective tape on the back of each shingle. Asphalt shingle manufacturers place a protective strip across the back of each shingle during the manufacturing process to keep the self-sealing feature on the fronts of the individual shingles from activating while the shingles remain in paperbound bundles.

Once shingles are installed on the roof in the standard overlapping *exposure* pattern, the protective tape will no longer prevent sunlight from activating the adhesive sealant. If a few protective tapes come loose while you are breaking apart bundles, pull whatever tape might come off easily and continue shingling.

Over-the-Top Shingling

A roof that has one layer of worn but not badly warped or curled shingles can be reroofed or *recovered* by *over-the-top shingling*. A new layer of shingles is installed directly on top of the worn shingles. A layer of new felt is not needed between the worn shingles and the new material.

Architectural or dimensional shingles—with an uneven texture designed to resemble wood shakes—cannot be re-covered with a second layer of shingles. The uneven surface layers of laminated shingles cannot be overlaid smoothly.

Use a utility knife to cut back the shingles along the rakes to expose about 3 inches of the deck so that just one layer of shingles is seen from the ground after the new roof is installed. As a result, the final appearance of the roof will be more attractive than if two layers could be seen at the rake.

It is also important to remove the tabs on only the third course of the old shingles. Remove only the tabs (Fig. 5-29); do not remove the entire course of shingles when you are shingling over-the-top. Done properly,

Fig. 5-29 Where you shingle over the top of one worn layer, tear off the tabs of the third course of the old shingles. Use a trowel to help pry off tabs.

removing the tabs will prevent the buildup of a hump along the third course when the new shingles are installed. If you do not remove the tabs at the third course, the hump will appear because of the layers of border shingles and the additional courses.

Take measurements and lay vertical chalk lines as described in chapter 4. Instead of laying horizontal chalk lines, it might be practical for you to *butt* the tops of the new shingles against the bottoms of the old shingles as you shingle the roof. This enables you to follow the pattern set by the old shingles, and it is an easy way to keep the new shingles straight. If you use this method, be certain that the proper 5-inch exposure is maintained and that the new shingles are precisely the same size as the old shingles.

The 45 Pattern

The instructions in this section are only for shingling square and rectangular roof sections. For guidelines on shingling angled roof sections, see the section "Starting Points" and chapter 6.

To begin nailing shingles using the *45 pattern* (Fig. 5-30), first refer to the section on laying vertical chalk lines in chapter 4. After you have snapped all the lines, take one bundle of shingles to the corner of the roof

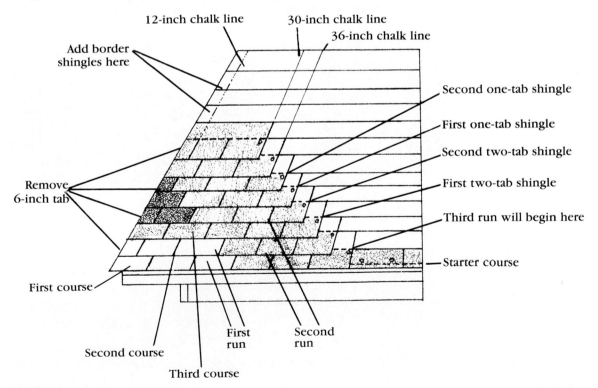

Fig. 5-30   The shaded shingles highlight the 45 pattern. Maintain the pattern by inserting one-tab shingles, two-tab shingles, and shingles with 6-inch tabs cut from the rake side.

Fig. 5-31 The first eaves' border shingle is installed so that it overhangs the drip edge by 1 inch and offsets the vertical 36-inch chalk line by 6 inches.

where you have snapped the 12-inch vertical lines. Nail the first eaves' border shingles and rake border shingles (Figs. 5-31 and 5-32).

Be sure to position the shingles so that the top, solid-colored portion of the shingles (granules up) faces outward (toward the ground) and the tabs face toward the roof. The border shingles must be aligned exactly with the chalk lines and be positioned so that they have a 1-inch overhang at the eaves and a 1-inch overhang at the rake.

➡ **Application Tip** It is best to get into the habit of nailing border shingles above the adhesive sealing strip. Otherwise you might easily drive nails where cutouts of the first course of shingles would expose the nail heads. This situation would lead to the development of leaks.

If you have installed the border shingles properly, there should be about 6 inches of *upside-down shingle* visible at the eaves so that the border shingle and the first-course shingle overlap rather than match up. A *match-up* is a common error that is easy to make and the result would be a series of leaks across the eaves.

First Course

With one vertical border shingle and one horizontal border shingle in place, begin the first course of shingles by positioning a full shingle so that it is

Fig. 5-32 Install the first rake border shingle so that it overhangs the drip edge by 1 inch at the rake and at the eaves. To show the proper alignment, the eaves' border shingle (Fig. 5-31) is not shown installed.

aligned with the 36-inch chalk line and extends over the rake by 1 inch. This first course goes directly on top of the border shingles (Fig. 5-33).

Second Course

The next step is to align the first shingle of the second course. Position a full shingle so that it extends 6 inches over the rake (Fig. 5-34). Carefully align the *factory edge* of the shingle with the 30-inch vertical chalk line. As you continue installing the pattern up the roof, the shingles will overlap so that a 6-inch tab extends over the rake.

➡ **Application Tip** Most professional roofers probably would let all the overhanging tabs remain, nail in place all the shingles on one section of the roof, then cut off all the tabs from the overhanging shingles at once. Figures 5-35 and 5-36 show how professional roofers use a chalk line to help them evenly trim a rake. Trimming rakes can be particularly difficult if you are unaccustomed to handling a utility knife. The results—especially with sunlight-baked shingles—can be considerable aggravation, scraped knuckles, and a rake that looks terrible. For do-it-yourselfers, there is an easier way to cut a rake.

Almost all weights and brands of three-tab asphalt fiberglass-based shingles have *factory cuts* at the 6-inch marks at the tops of the shingles.

Fig. 5-33 The first course is installed aligned with the 36-inch chalk line and even with the border shingles.

Fig. 5-34 Align a full shingle with the 30-inch chalk line. Align the bottom of the second-course shingle with the tops of the first-course watermarks. Trim the 6-inch overhanging tab before or after the shingle is installed.

Fig. 5-35 (Left) Use a chalk line to mark the proper 1-inch overhang.

Fig. 5-36 (Top) Trimming the rake with a utility knife is a difficult task for a beginner. Use tin snips for this job.

This is a very convenient and accurate starting point for cutting off the tabs. While the easiest tool with which to make cuts is a pair of tin snips, professional roofers use a utility knife with a hook blade or a straight blade because it is faster. Many roofers prefer to use hook blades to cut asphalt fiberglass-based shingles because straight blades tend to "slide" over the shingle. The hook blades will "grab" the shingle and give a better cutting action.

Whether you choose to use tin snips or a utility knife, first temporarily position but do not nail the shingle in place. Next, mark the shingle where it is to be cut, then turn over the shingle so that the granules face down. Find the factory cut on the proper end of the shingle and cut off the tab. Be careful not to cut all the way through a shingle and into whatever is underneath, and take care not to cut off the wrong end of the shingle. If you make a wrong cut, put the scraps aside; you'll undoubtedly be able to use the scraps at the other end of the roof.

Remember to first position the shingle, and then trim the overhanging

Fig. 5-37 Cut a one-tab for later use.

Fig. 5-38 Align the first shingle of the third course.

tabs as you install the course of shingles along the rake. If you use this method, mistakes are less troublesome to correct.

Third Course

Position the shingle (minus the 6-inch scrap) so that it aligns with the 30-inch vertical chalk line, the rake (with a 1-inch overhang), and the horizontal chalk line. Nail the shingle in place.

The next step is to take two full shingles and cut off $1/3$ of the tabs—so-called *one-tab shingles*—from the rake side of both shingles (Fig. 5-37). Save the $1/3$ tabs for use during the next run. The reason for saving the tabs is explained on page 95.

Fourth Course

Begin the fourth course of shingles by positioning one of the $2/3$ shingles with the cut end along the rake. The factory edge must be aligned with the 6-inch factory cut of the shingle you have nailed in place for the second course (see Fig. 5-38).

Fifth Course

Cut the 6-inch tab from the rake side of the second $^2/_3$ shingle, and carefully align and position it for the fifth course. You should be able to clearly see the shingles taking on a 45-degree pattern of "steps" up the roof. The cutouts must all be aligned in the proper 6-inch pattern. It is essential that these first few courses be positioned properly. An error can easily be fixed at this stage. If you are not absolutely sure that the pattern is correct, stop and make the necessary adjustments.

Remember to install border shingles (Fig. 5-39) and then begin the next run of the pattern (Fig. 5-40). By nailing a border shingle and four full shingles in place, you will reach the top of the pattern *run*. To extend the stepped pattern further up the roof, nail a full shingle even with the rake, cut a 6-inch scrap piece from the rake side of another full shingle (save the scrap for later), and nail the shingle in place.

Fig. 5-39 Install another border shingle along the rake.

Fig. 5-40 Remove the tabs of the third course of the worn shingles and install a full shingle.

At this point, retrieve the one-tabs you saved from the first run. Nail the two one-tabs in place to retain the step pattern of the shingles. Continue the pattern by cutting *two-tab shingles* and 6-inch scraps from the appropriate shingles, as you did for the first run. Repeat this pattern on the remaining runs (see Figs. 5-41 through 5-48).

The Random-Spacing Pattern

The techniques for applying asphalt fiberglass-based shingles in the *random-spacing pattern* are almost identical to the methods used for shingling with the 45 pattern. With the random-spacing pattern, the cutouts on the shingles align every sixth shingle. With the 45 pattern and the straight pattern, shingle cutouts help reduce wear patterns from rain as the shingles age. Consequently, the shingles will last slightly longer than identical shingles installed in the 45 pattern or the straight pattern. Shingles with no cutouts or shingles manufactured to metric dimensions are ideal for installation using the random-spacing pattern.

Begin the random-spacing pattern by snapping chalk lines as described in chapter 4. After you have snapped all the lines, take a bundle of shingles to the corner of the roof where you have snapped the 12-inch vertical lines. Nail the first vertical border shingle. As with the 45 pattern, be sure to position the shingle so that the top, solid-colored portion of the shingle faces outward and the tabs face toward the roof.

The border shingle must be aligned exactly with the chalk line and be positioned so that it has a 1-inch overhang at the bottom and a 1-inch overhang at the rake. Now position a horizontal border shingle. Turn this shingle so that the granule-surfaced portion of the tabs face you. Align the shingle exactly with the horizontal chalk line and nail the shingle in place.

Nail the border shingles above the adhesive sealing strips on the shingles. Do not drive a nail where a cutout of the first course of shingles would expose nail heads, causing a leak.

First Course

The difference between the 45 pattern and the random-spacing pattern begins with the first course. Instead of beginning the first course of shingles by positioning a full shingle so that it is aligned with the 36-inch chalk line and extends 1 inch over the rake, position the first-course shingle of the random-spacing pattern so that it extends 3 inches over the rake. Nail the shingle in place and trim the overhanging piece even with the border shingle (leaving a 1-inch overhang).

Second Course

Align the first shingle of the second course by positioning it so that 9 inches of the shingle overhangs the rake and the other side of the shingle aligns

Fig. 5-41 (Above left) Install the second shingle of the third course.

Fig. 5-42 (Above right) Tabs are easier to cut when the slots face away from you.

Fig. 5-43 (Left) Installing the first shingle of the fifth course.

Fig. 5-44 Cut two-tab shingles and install them on the second run to continue the 45 pattern.

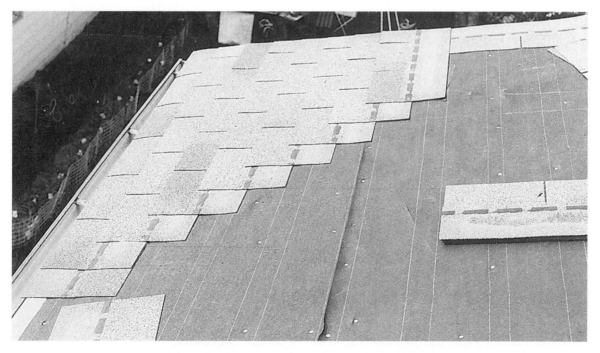

Fig. 5-45 Two runs of shingles have been installed.

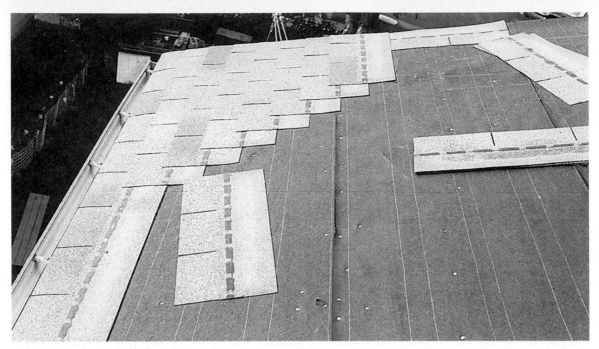

Fig. 5-46 The border shingle and the first shingle of the third course are installed. Continue the pattern by installing full shingles.

Fig. 5-47 With the 45 pattern, the factory cutouts must be aligned on every other course.

Fig. 5-48 A few more runs will take the pattern to the ridge.

with the 6-inch factory cut on the first-course shingle. Trim the overhanging piece.

The 30-inch and the 36-inch chalk lines will not intersect the shingles until several courses in the random-spacing pattern have been installed. Be careful not to match up the bottom border shingles and the first course.

Third Course

Begin the third course of shingles by positioning a two-tab shingle with a portion of the shingle overhanging the rake and the other end carefully aligned with the 6-inch factory cut of the second-course shingle. In other words, the third course must be placed so that it continues the 45-degree step pattern.

Fourth Course

Position another two-tab shingle for the fourth course by aligning it with the 6-inch factory cut of the third-course shingle; continue the step pattern. Trim the overhanging piece.

Fifth Course

Install border shingles and begin a run of full shingles from the bottom. When you reach the top of the pattern, install one-tabs to extend the stepped pattern.

The Straight Pattern

Nailing shingles in the *straight pattern* aligns the cutouts 6 inches apart on every other shingle. It is best to use this pattern on steep surfaces that require the use of roof jacks (Fig. 5-49), on roof dormers, and on short, vertical surfaces such as catwalks. The pattern consists of repeating the vertical runs up the roof until a section of the roof has been completely covered.

Begin the straight pattern by laying chalk lines as described in chapter 4. After snapping all the chalk lines, take a bundle of shingles to the corner of the roof where you have snapped the 12-inch vertical lines. Nail in place the first horizontal border shingle (Fig. 5-31). Be certain to position the shingle so that the top, solid-colored portion of the shingle (granules up) faces outward (toward the ground) and the tabs face toward the roof. The border shingle must be aligned exactly with the chalk line and be positioned so that it has a 1-inch overhang at the bottom and a 1-inch overhang at the rake.

Position a vertical border shingle, then turn this shingle so that the tabs face you. Align the shingle exactly with the horizontal chalk line, and nail the shingle in place (Fig. 5-32).

Fig. 5-49 Use the straight pattern on steep roofs that require roof jacks.

➡ **Application Tip** Make a habit of nailing the border shingles above the adhesive sealing strip. Otherwise, you might easily drive a nail where a cutout of the first course of shingles would expose the nail head. Exposed nail heads cause leaks.

First Course

With one vertical and one horizontal border shingle in place (Fig. 5-50), begin the first course of shingles by positioning a full shingle so that it is aligned with the 30-inch chalk line and extends over the rake. This first course goes directly on top of the border shingles (Fig. 5-51). If you have installed the border shingles properly, there should be about 6 inches of upside-down shingle visible on the outside of the vertical chalk lines (Fig. 5-50) so that the border shingle and the first-course shingle overlap rather than match up. A match-up is a common error that is easy to make, and the result would be a series of leaks across the eaves.

Second Course

The next step is to align the first shingle of the second course. Position a full shingle so that it is even with the rake. Carefully align the factory edge of the shingle with the 36-inch vertical chalk line. Now you have a choice of cutting off the 6-inch scrap or temporarily leaving the overhanging tab in place until the run has been completed. Most professional roofers would probably let the overhanging tabs remain in place and then cut off all of them at once.

Cutting overhanging tabs can be particularly difficult for someone unaccustomed to handling a utility knife; it's easier to use a pair of tin snips.

Fig. 5-50 Rake and eaves border shingles are installed for the straight pattern.

Fig. 5-51   Install the first shingle in the first course of the straight pattern so that a 6-inch tab overhangs the rake.

If you do prefer to use a utility knife, a hook blade is easier to use than a straight blade. Trim the second-course shingle so that it is even with the vertical border shingle.

Third Course

Begin the third course of shingles by positioning a full shingle so that it is aligned with the 30-inch chalk line and extends over the rake. This shingle must be aligned so that the bottom of it is even with the cutouts of the second-course shingle (see Fig. 5-52).

Continue the straight pattern by installing shingles in the vertical run until you reach the top of the section (Fig. 5-53). Every other shingle will overlap the rake by 6 inches and the alternating full shingles will align with the border shingles at the rake. A shingle can be used as a straightedge (Fig. 5-54) to help you trim the 6-inch tab from the second shingle in the *starter course*. Use the factory cut as a guide.

Be absolutely certain to nail each shingle with four nails. To do this when installing the straight pattern, you must lift the tabs on every other shingle course as you install each of the runs following the first run.

➡ **Application Tip** While you are installing shingles in the straight pattern, you can save time and effort by using the following techniques for positioning bundles of shingles. If the bundles of shingles do not slide as a result of the steepness of your roof, place a bundle on the run of shingles you have just installed. The way you position the bundle is crucial to obtaining the greatest shingle application efficiency. Position the new bundle of shingles with the tabs facing "up" and the cutouts facing toward the deck. The bundle must be close enough to reach, but it must not prevent you from lifting the tabs of the run you have just installed.

To position a partial bundle of shingles that otherwise might slide away on a slightly steep deck, you can temporarily insert a piece of step flashing under one of the shingle tabs of the straight-pattern run just installed. Lift a tab and gently slide one edge of two or three L-shaped pieces of step flashing snug against the nail holding the shingle in place. The idea is to create a temporary resting place for the loose shingles that are about to be installed. Again, the shingles must be within reach but they must not prevent you from lifting the shingle tabs as you install the current run of the straight pattern. When you are finished with the run, be careful not to damage the shingle tab as you remove the step flashing.

Fig. 5-52 Align the shingles with the 30- and 36-inch vertical chalk lines. Also, align the bottoms of the shingles with the watermarks of the course below.

Fig. 5-53 Four courses of the straight pattern have been installed.

Fig. 5-54 To cut 6-inch tabs from the alternating shingles in the starter course, use a full shingle as a straightedge.

Capping

Capping can be cut into square or diamond shapes. Most roofers will cut caps into diamond shapes if appearance is important (such as it can be seen from the ground); otherwise, capping can be cut into squares. Hips and ridges are shingled with capping. If your home has hip sections, always finish capping the hips before you cap the ridges. Determine the number of caps you will need by referring to the section on estimating materials in chapter 3.

The fastest way to cut capping from bundles of three-tab, asphalt fiberglass-based shingles is to first turn the bundles granular-side down and tear away the wrapping paper. Cut only one bundle of capping at a time in case you have overestimated the amount of capping you will need.

Work on your knees and position the cutouts facing away from you. Begin at either side of the bundle. Steady the top shingle with the palm of your hand that is not holding the utility knife. Using a hook-blade utility knife, cut off a diagonal slice of shingle beginning at the watermark and going all the way to the top of the shingle (Figs. 5-55 and 5-56).

Work slowly with a steady pulling pressure on the knife until you get the hang of cutting. You should soon be able to make steady, successive slicing motions. Always be very careful not to scrape your knuckles. Don't jerk the knife while you attempt to make cuts.

Make the next cut starting at the second watermark. Angle the cut so that you are slicing toward your body. The object is to form *diamond-shaped capping.* Cut the remaining two sides from the shingle by slicing the scrap from the shingle. Cut the diamond shape for all the caps. Diamond-shaped caps are much more attractive than square-cut capping.

➡ **Application Tip**   Here is a simple and accurate way to lay capping. Use a chalk line and three caps to determine if you have the proper coverage. Position one cap at each end of the roof and place one cap at the center of the hip or ridge. Position the center and end caps so that they straddle the hip or ridge and cover the watermarks on both the side courses and the top courses.

If you are working alone, tap in a nail at the bottom corner of the street side of the center cap, hook your chalk line on the nail head, and walk to either end of the roof. If you are working with others, instead of using a nail, have someone hold one end of the line in position.

Align the chalk line with the bottom of the end cap. Snap a line, rewind the chalk line, snap another line at the other end of the roof (aligned with the other cap), and check to see if the line is below all the 5-inch watermarks on the side courses or the top courses of shingles. If the chalk line shows that you will have adequate coverage, you can begin nailing caps. Otherwise, you will have to install another course of shingles. For easier installation, bend but do not crease the caps before positioning them.

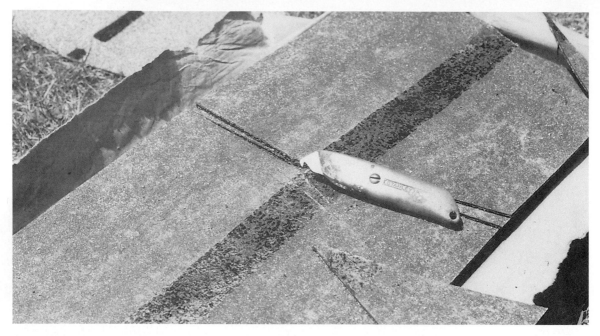

Fig. 5-55 To cut capping from full shingles, first turn the bundle face down, then use a hook-blade utility knife or tin snips to cut diagonal scrapes from the watermarks to the top of the shingle.

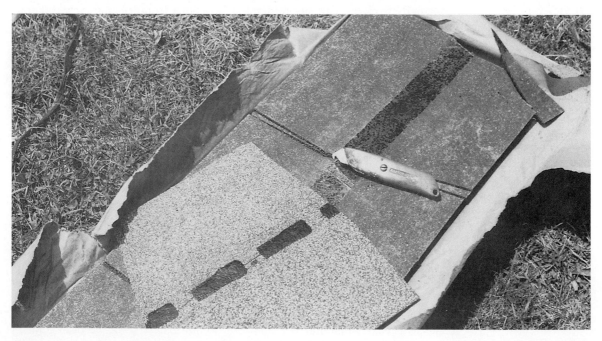

Fig. 5-56 Trim both sides of the one-tabs to make diamond-shaped capping.

Hip Capping

Because the last hip caps should go under the ridge caps, cap all hips before you install ridge capping. Before you install hip capping, make certain that all the shingle courses on both roof sections adequately cover the hip. Snap a chalk line using the techniques described in the preceding section.

When installing the first hip cap at the eaves, position the cap so that the bottom center of the granular portion is even with the 1-inch overhang at the eaves. You will find that the edges of the cap overlap the bottom shingle course on both sections of the roof. Trim this portion of the cap at an angle so that the cap is even with the bottom courses. Install the caps on the hip by aligning the granular side even with the first cap's 5-inch watermark and the chalk line. Drive a nail into the cap about 1 inch up from the chalk line and between the adhesive strip on the cap and the 5-inch watermark. Drive another nail at the opposite location on the other side of the hip.

Install the second cap so that it aligns with the chalk line and overlaps the first cap. Each additional cap should provide 5 inches of coverage. The caps must overlap so that each cap is double nailed.

After you have nailed several caps, you will have room to turn your back to the roof edge and sit facing the ridge. If you are working alone, hold about 20 caps in your lap and work up the hip. If help is available, have someone hold each cap in place, even with the chalk line, while you nail. Your helper should sit facing you.

When you reach the end of the hip, cut the last cap so that it "folds" over the ridge. Cover any exposed, *face-nailed* fasteners with roofing cement.

Ridge Capping

Before you install the ridge caps, make sure that the top courses of shingles provide adequate coverage on both sides of the ridge. The caps must cover the 5-inch watermarks on both top courses across the entire ridge (see Figs. 5-57 and 5-58).

The traditional procedure to install ridge caps is to place the unnailed ends of the caps opposite the prevailing winds. Begin at one end of the roof rather than in the middle.

Align the first cap on the ridge with the granular side even with the 1-inch overhang (Fig. 5-58) and the chalk line. Drive a nail into the cap about 1 inch up from the chalk line and between the adhesive strip on the cap and the 5-inch watermark. Drive another nail at the opposite location on the other side of the ridge.

Install the second cap so that it aligns with the chalk line and overlaps the first cap. Each additional cap should provide 5 inches of coverage. The caps must overlap so that each cap is double nailed (see Figs. 5-59 through 5-63).

When you reach the end of the ridge, trim the last cap even with the 1-inch overhang so that it shows only the granular portion of the cap. The last

Fig. 5-57   Trim the top of the last course of shingles. Overlap the tops of the opposite, last course and nail or trim.

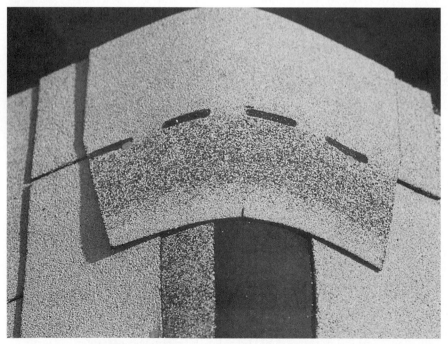

Fig. 5-58   Align the first cap along the ridge.

Fig. 5-59 Overlap the caps so that they can be double nailed.

Fig. 5-60 Align and carefully nail the edge of the cap.

Fig. 5-61 Center the caps at the ridge, overlap each cap by 5 inches, and drive one nail on each side about 1 inch from the edge of the cap.

Fig. 5-62 Continue installing capping until you reach the opposite rakes.

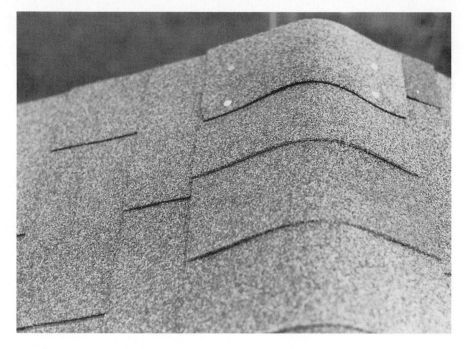

Fig. 5-63 Trim the last cap to align with the rakes and drive four nails to secure the corners. Add a dab of roof cement on the top of the four nail heads.

cap most likely will be less than a full cap. Drive four nails into the last cap and cover exposed nail heads with roofing cement.

If the roof has hips, the first and last ridge caps must be cut so that they overlap the hip sections. Cut the end caps so that they "fold" over the hip sections. Cover any exposed nail heads with roofing cement.

Dormer Ridge Capping

To install caps on a dormer ridge, follow the instructions outlined in the preceding sections. When you reach the valley section of the ridge, make certain that the last few caps go under the shingles at the top of the valley.

Be extremely careful when you nail the caps as you approach the valley. Use common sense to judge where the water will run and where it is safe to nail the caps. Never drive a nail into the exposed metal portion of a valley.

Chapter **6**

Angled Roof
Sections

Shingling rectangular and square sections of a roof is a matter of selecting a pattern, laying chalk lines, and repeating the shingle pattern until the sections are completed. When you install three-tab asphalt fiberglass-based shingles on odd-shaped, extremely steep, or angled roof sections such as hips, wings, valleys, dormers, and mansards, careful planning is required.

Shingling each type of roof section described in this chapter requires a technique called *filling in*, or *backing in*. Filling in a roof section usually means striking two parallel chalk lines at a 6-inch width to square off an angled section of roof in order to get the longest possible vertical run (the straight pattern) or stepped run (the 45 pattern or the random-spacing pattern) of shingles. With the random-spacing pattern, it is possible to strike one 45-degree chalk line as guidance for applying the shingle pattern.

For speed and accuracy, most shingling on angled roof sections should be completed with the work in front of you—right to left if you are right-handed or left to right if you are left-handed. The remaining portion of the roof section is then filled in by working "backwards" across the roof; in other words, from the chalk lines toward the rake.

Figures 6-1 through 6-3 show how the application pattern with the standard 5-inch shingle exposure is maintained past an angle in the roof line. First, border shingles are installed at the rake. Next, the pattern of shingles is filled in by first *high-nailing* the courses as they are backed into position. The shingles are then nailed properly once the backed-in shingle courses are evenly placed with the adjacent courses having the standard 5-inch exposure at the watermarks. A shingle or a course of shingles can be high-nailed by driving nails about an inch from the top edge of each shingle. After this temporary pattern is completed, all backed-in courses must be nailed properly with four nails in each full shingle.

Fig. 6-1 Install border shingles around a break in the roof section.

Fig. 6-2 The courses are continued with the standard exposure at the watermarks by first high-nailing the intersecting shingles.

Fig. 6-3 The 45 pattern is maintained past the break in the roofline.

To maintain the standard shingle pattern beyond a sharp angle or cut in the roofline, it is sometimes necessary to install a *short course* of shingles with less than the standard 5-inch exposure. The result is one short course across the remainder of the roof section beginning where the roof angle or cut intersects.

As you approach the angle (Fig. 6-4) or cut in the roofline, the border shingles along the eaves are installed (Figs. 6-5 and 6-6), the rake shingles (Fig. 6-7) are trimmed and installed along the rake, and the pattern is maintained (Figs. 6-8 through 6-12) until the courses reach the top of the angle.

Where the angle intersects the two sections (Fig. 6-13), a short course (Fig. 6-14) is installed until the rake is reached.

Hips

Any of the three patterns described in chapter 5 can be used to shingle a *hip roof*. The first step is to lay horizontal chalk lines, as described in chapter 4. Because hip roofs have no rakes, carefully determine where you mark the vertical chalk lines.

If you are working on a new roof or if you have torn off the old shingles, find a rafter at the top left-hand corner of the hip (if you are right-handed) and use it as a reference point for the vertical chalk lines (see Fig. 6-15). Mark the vertical lines by following the basic instructions outlined in chapter 4.

Begin shingling the remaining angled portion of the hip sections. As you approach the hip with each course of shingles, use a utility knife or tin

Fig. 6-4   Drip edge is installed along the eaves and rake of the angle in the roofline.

snips to trim the shingles at an angle. The last shingle in each course must provide enough coverage so that the capping overlaps it.

As you complete the runs, fewer full shingles are required. Use two-tab, one-tab, and smaller shingle portions to fill in the section. Keep the cutouts aligned. Shingling a hip requires a great deal of cutting and trim work (Fig. 6-16). Be careful!

Wings

A *wing* section of a roof can be shingled much like a hip section. First, snap horizontal chalk lines as outlined in chapter 4. Square off the angled section to allow for the longest possible vertical run of shingles (see Fig. 6-17).

Fig. 6-5 Border shingles are installed at the angle.

Fig. 6-6 The shingle for the first course is positioned and then marked to be trimmed with a utility knife.

Fig. 6-7 The marked shingle is turned over and then cut from the back.

Fig. 6-8 The first course is nailed in place.

Fig. 6-9   The second course is positioned.

Fig. 6-10   The third course is trimmed with a 1-inch overhang.

Fig. 6-11   The 45 pattern is maintained.

Fig. 6-12   The 45 pattern reaches the angle in the roofline.

Fig. 6-13 After additional border shingles are installed, the next course of shingles is a short course that intersects the angle in the roofline.

Fig. 6-14 The short course provides about 2 inches of exposure but maintains the application pattern until the opposite rake is reached.

If you are reroofing over a layer of worn shingles, mark vertical chalk lines 6 inches apart using the cutouts of the old shingles as guidelines. If you are shingling new work, be extremely careful when you mark off the measurements. Use a rafter at the top of the wing as one reference point and the bottom corner of the wing as the other reference point for the 6-inch-wide vertical chalk lines.

Follow the basic instruction for laying vertical chalk lines given in chapter 4, and make adjustments for the angled rake wing.

Before you begin to nail shingles, be absolutely certain that the measurements and chalk lines are correct. Shingle the squared-off portion of the wing first, then fill in the remaining angled section.

Valleys

A roof *valley* is formed where two sloping roof sections intersect to create an internal roof angle or channel that drains runoff and meltwater from

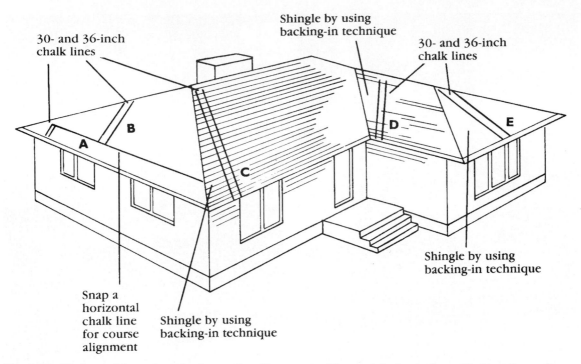

Fig. 6-15 Work on a hip roof section by section. Keep most of the work in front of you. Find a rafter and snap chalk lines so that the bottom third of the hip section can be shingled (A). After the bottom third of the roof has been shingled, snap chalk lines and shingle the top section (B). Square off this hip-roof section by snapping parallel 6-inch-wide chalk lines (C). Square off this section and back the shingles into the valley area (D). Shingle small hip-roof sections by snapping chalk lines at the center of the section (E).

higher roof sections. The surface of the valley channel must be reinforced with galvanized metal (Figs. 6-18, 6-19, and 6-52) or copper, roll roofing (Figs. 8-1 and 8-17 through 8-19) or other mineral-surface membrane, or shingles specially installed in an overlapping *Western weave valley* pattern. With a Western weave, shingle courses extend fully across one side of the angle of a valley and the opposite course from the other angled roof section. The shingles are overlapped, then trimmed at the center line of the valley.

When you measure the length of the valleys for metal or membrane, add an additional 8 inches for each overlap of valley material at the ridge and eaves. If you cannot obtain metal or roll roofing in a long enough piece to cover the valley length, add another 6 inches for each overlap between sections of valley material. Install a double layer when you use mineral-surface roll roofing. All valley material must be at least 36 inches wide. See Figs. 6-20 through 6-24.

Before you install the valley material, install drip edge and border shingles at the eaves. See the section on application patterns in chapter 5. Shingle all of the roof sections until just before the runs reach the valley. The bottom section of the valley material must go on top of the border shingles at the eaves on both adjoining sections of the roof at the bottom of the valley.

After the drip edge, border shingles, and first-course shingles are in place, center the bottom section of the valley material where the roof sections are joined. The center of the valley material should extend no more than 1 inch over the bottom of the border shingle. Several inches of the edge of the valley material will extend over the border shingles when it is positioned. Nail the first section of valley material in place, then use tin snips to trim the excess overhang.

Fig. 6-16 Trim shingle courses that overlap a hip with a utility knife or aviation snips.

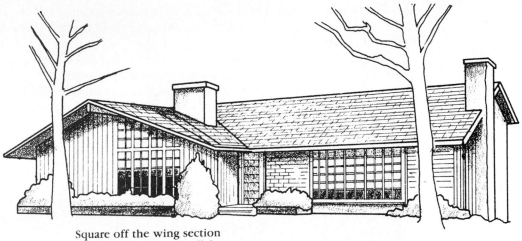

Square off the wing section
with 6-inch-wide, parallel
chalk lines

Fig. 6-17 A wing section can be squared off for shingling.

Never drive a nail closer than 3 inches from the edge of the valley material. Remember that a great deal of water will be channeled over the valley surface. The easiest way to nail down valley material is to position it, make certain it is centered, and then drive nails every 2 feet or so along one edge of the metal or mineral-surface roofing. After you have secured one side of the valley section, drive nails on the other side—

Fig. 6-18 Galvanized metal is a common valley material.

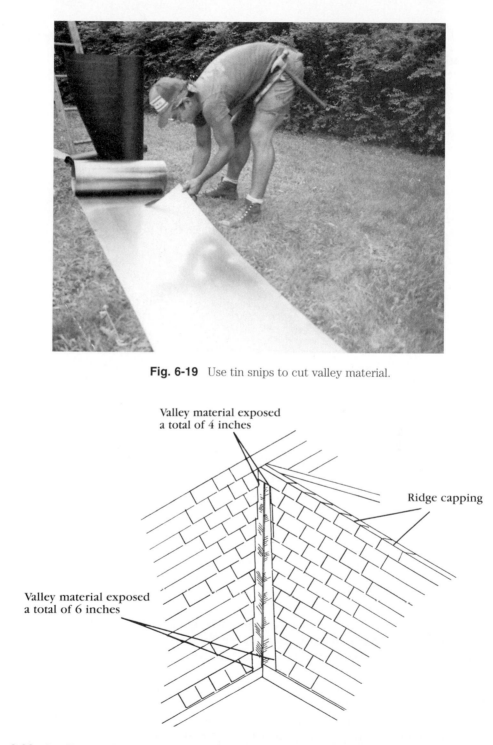

Fig. 6-19 Use tin snips to cut valley material.

Valley material exposed
a total of 4 inches

Ridge capping

Valley material exposed
a total of 6 inches

Fig. 6-20 A valley can be trimmed so that the top is narrower than the bottom.

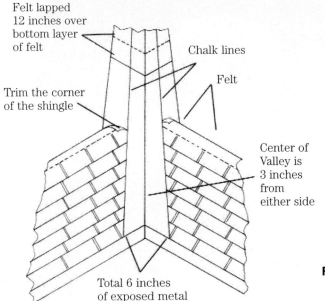

Felt lapped
12 inches over
bottom layer
of felt

Chalk lines

Felt

Trim the corner
of the shingle

Center of
Valley is
3 inches
from
either side

Total 6 inches
of exposed metal

Fig. 6-21   An open valley with metal flashing.

about every 2 feet—at staggered locations from the other nails rather than opposite them.

Where sections of valley material overlap, provide at least 8 inches of double coverage. Be absolutely certain to place the top overlapping section above the other valley material section.

At the ridge line, use tin snips to cut the valley material so that it overlaps the intersection of the valley by 8 inches. When you install the other side of the valley, make sure that the valley material again overlaps the ridge lines. Nail carefully at these points.

When valley material is in place, install border shingles along the length of the valley. Border shingles provide additional protection from water damage and help channel the runoff. Border shingles also provide a straight line on both sides of the center of the valley that prove quite handy when trimming the scrap from each course of shingles. Take your time when you trim the valley shingles. A neat job looks very attractive; an improperly trimmed valley is an eyesore.

To install border shingles along a valley, first find the center of the valley at the eaves. Measure 3 inches from the center on both sides (for a total width of 6 inches). Lightly scratch a set of Vs at two, 3-inch measurement points (at the tops and bottoms of the border shingles). Position full shingles with the granular side up and the cutouts facing away from the center of the valley.

The shingle tops must be aligned with the measurement marks. Check to determine if you have a full 6 inches of exposed *open valley* between the two opposing border shingles. Drive two or three nails in the granular

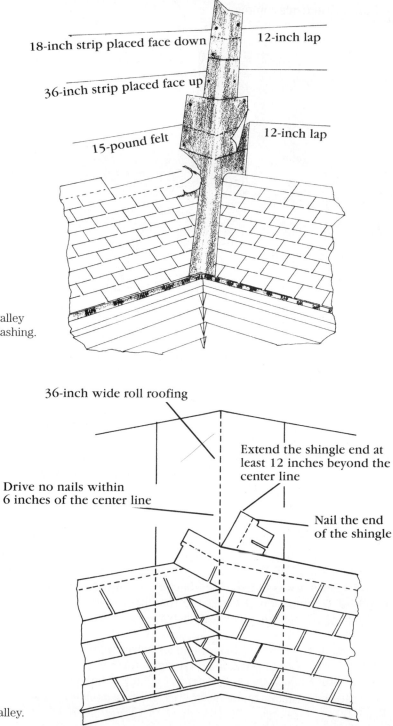

Fig. 6-22 An open valley with roll roofing for flashing.

Fig. 6-23 A woven valley.

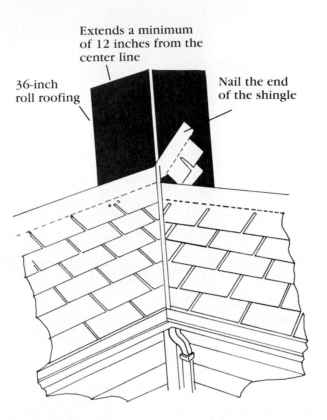

Extends a minimum
of 12 inches from the
center line

36-inch
roll roofing

Nail the end
of the shingle

Fig. 6-24 A closed-cut valley.

portion of the tabs of each shingle to hold them in place. Do not drive nails near the exposed valley material.

At the ridge line, install two more border shingles. Snap chalk lines at the tab sides of the border shingles (away from the middle of the valley) to serve as guidelines, then fill in the remainder of the valley borders with shingles. You can now resume roofing the sections by following the application pattern.

Using the Western weave to shingle a *closed valley* (Figs. 6-25 through 6-35) without an underlayment of metal or roll roofing is common in regions with low annual rainfall. Where rain is frequent and snow and ice accumulate, metal valleys are generally used because of the likelihood of substantial runoffs. A combination of either roll roofing or metal installed under weaved shingles can provide an especially watertight valley for very wet climates.

As you shingle toward a valley or weave shingles across a *laced valley* you will often find it convenient to position one shingle at a time, cut it, and then nail it in place. Other times, it is necessary to extend full shingles into the valley (where the ends will be trimmed) by first inserting a one- or two-tab shingle (Fig. 6-29).

Fig. 6-25 At least one and preferably two full shingles must be positioned under the intersecting valley shingles of the roof section above to ensure a watertight valley.

Fig. 6-26 The left-hand roof section's eaves' border shingle extends across the valley.

Fig. 6-27   The starter course extends across the intersecting valley shingles of the roof's right-hand section.

Fig. 6-28   The second course of the random-spacing pattern for the left-hand side of the valley.

Fig. 6-29 Insert a one- or two-tab shingle into the pattern to extend a full shingle across the center of a valley.

Avoid driving any nails into the center of the valley material. Be extremely careful not to cut into the valley material or shingles. Any cut, punctured, or torn valley sections must be replaced with watertight valley material.

Snap a chalk line down the center of the valley to indicate where to trim the edges of the overlapping weaved shingles and to align the random-spacing pattern (Figs. 6-36 and 6-37). This alignment technique can be used only with the random-spacing pattern. Shingles applied using the 6-inch offset pattern must be filled in from 30- and 36-inch parallel vertical lines (Fig. 6-15).

Where a valley intersects below the ridge line of the main roof, high-nail a course of shingles so that the pattern reaches the rake (see Figs. 6-38 through 6-43).

Fig. 6-30   The left-hand roof section is shingled across the valley.

Fig. 6-31   Continue to shingle the left-hand valley section across the valley and toward the ridge.

Fig. 6-32 The random-spacing pattern is extended toward the ridge.

Fig. 6-33 Full shingles extend across the valley.

Fig. 6-34   The top course extends over the intersection of the ridge and valleys.

Fig. 6-35   With the majority of the left-hand section of the valley already installed, the right-hand side of the valley is woven over it by installing and trimming shingles in the 45-degree, random-spacing pattern.

Fig. 6-36 (Right) Snap a chalk line down the center of the valley.

Fig. 6-37 (Left) The random spacing pattern minimizes cut work in a valley and maximizes the lengths of shingle runs in the application pattern.

Fig. 6-38 (Right) The random-spacing pattern, with a 5-inch exposure, is continued toward the opposite rake by high-nailing one course where the pattern intersects with the ridge line and the valleys. Remember that ridge capping must be inserted under the overlapping full shingle.

Fig. 6-39   The pattern is backed in toward the rake.

Fig. 6-40   Continue filling in shingles at the rake.

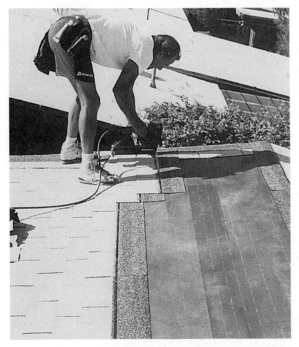

Fig. 6-41   The 45-degree, random-spacing pattern is resumed.

Fig. 6-42   Additional tie-in runs are made toward the ridge.

Fig. 6-43 As tie-in courses are completed, full runs from the eaves to the ridge can be installed.

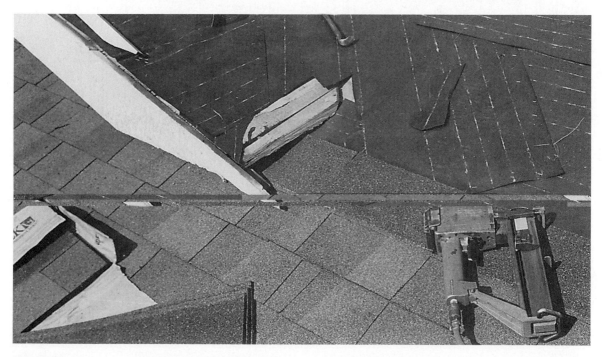

Fig. 6-44 Where a valley channel empties onto the main roof, field shingles must be installed under the valley shingles. Note the damaged plywood to be replaced.

When you shingle at the bottom of a dormer valley (Figs. 6-44 through 6-48), check that the border shingle and the first-course shingle directly below the dormer valley are both under the valley material or under the first intersecting Western-weave shingle.

As you weave courses of shingles across a valley, trim each shingle in the pattern and remove the scrap along the valley channel or at the valley border shingle. Use tin snips or a utility knife with a hook blade to trim the scrap from each shingle as you nail the courses in place. See Figs. 6-49 through 6-51.

Dormers

Shingling the tops of *dormers* generally consists of covering two short sections that taper into valleys. Measure for and snap chalk lines as described in chapter 4. Any of the three shingle application patterns can be used to cover a dormer, but the random-spacing pattern or the straight pattern will probably be the easiest methods.

Fig. 6-45 After a small section of damaged plywood was replaced, the rake border shingle was installed and the first partial course of the woven valley shingles positioned.

Fig. 6-46 The second partial course is installed at the intersection of the rake and valley.

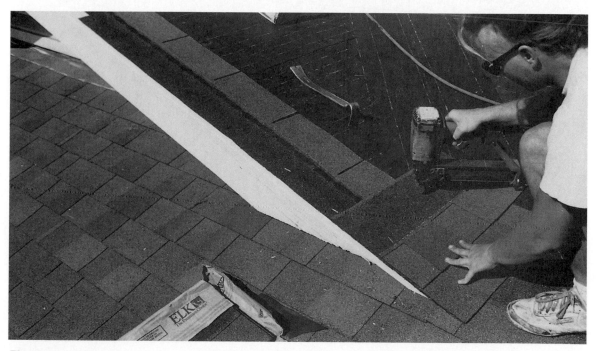

Fig. 6-47 A third partial course is installed to provide additional protection against runoff where the rake and valley meet.

Begin shingling a dormer at the rakes. You can complete the first runs at the rakes and place a few bundles of shingles within easy reach of the remaining work area.

On one side of the dormer, you will be able to shingle with the work in front of you (Fig. 6-52). On the other side of the dormer, start at the rake and back in the shingles from the rake. Because a dormer usually can be shingled with a few bundles, it is not practical to square off the angled valley section as you would for hip or wing sections.

To properly shingle the main section of a roof that has a dormer, you must carefully plan the shingle pattern so that the shingle courses will match past the top of the dormer and on the other side of the dormer (see Fig. 6-53).

To tie in the courses and keep the shingle cutouts aligned, shingle the rake side of the dormer—where you first began to nail shingles—past the top of the dormer ridge. Also shingle the section of the main roof directly below the dormer and all the way to the other rake.

On the first rake side of the dormer, measure the distance from the bottom course of shingles, including the 1-inch overhang at the border shingle, to the top of the course of shingles immediately above where the dormer

Fig. 6-48 The random-spacing shingle pattern allows the installation of shingles to extend from the field into and across the valley.

Fig. 6-49 Shingles can be woven across a valley using a chalk line for guidance for overlapping courses.

Fig. 6-50 Additional shingle courses are installed across the dormer valley.

Fig. 6-51 No-cutout, laminated shingles installed using the 45-degree, random-spacing pattern and the Western-weave valley pattern.

ridge line and the valley meet (30 feet, for example). At the other side of the dormer, scratch a V at the same distance up from the border shingle (for example, 30 feet).

Strike a horizontal chalk line to guide the placement of the *tie-in course* of shingles. The tops of the tie-in course must touch this horizontal line. Position the tie-in course by high-nailing two nails in each shingle no lower than 1 inch from the horizontal chalk line. These shingles will be nailed in later in the standard locations above the cutouts and below the adhesive strips. There is no need to remove the high nails.

After you have installed the tie-in course past the dormer, strike two parallel chalk lines 6 inches apart, as close to the dormer as possible. The parallel lines must match the cutouts of the tie-in course and the cutouts of the shingles below and to the side of the dormer. Be extremely careful not to break the proper shingle-pattern sequence. Use the straight pattern to make one run up the parallel lines. Make certain that the application sequence is correct!

Fill in the sections under the tie-in course by lifting the high-nailed shingles and slipping the next course of shingles into position one at a time. Nail the tie-in shingles above the watermarks and below the adhesive strips. Remember to drive four nails into each shingle. Continue *backing down* the courses until you have finished the section.

Fig. 6-52 On the left side of the house, the roof can be shingled from the rake into the valley. On the dormer, back in the shingle pattern from the left rake of the dormer.

Mansards

Because you must work from a ladder or a scaffold, *a mansard roof* can be the most challenging type of surface to shingle for do-it-yourselfers and professionals. If you do not have the proper equipment or the confidence to do a first-rate job—that means safely getting all the shingles on straight in a reasonable amount of time—hire a professional roofer to do the mansard portion of the work.

As with any roof surface, you can use roofing nails or staples to apply felt paper to the deck of a mansard roof. A hammer tacker (Fig. 2-6) allows building paper to be easily and rapidly stapled in place. Use a utility knife to trim felt around windows and at the rakes.

Frequently, mansard roofs are designed with one or more windows. Step flashing or specially designed channel flashing must be installed where the shingles and windows meet. Be sure to discuss

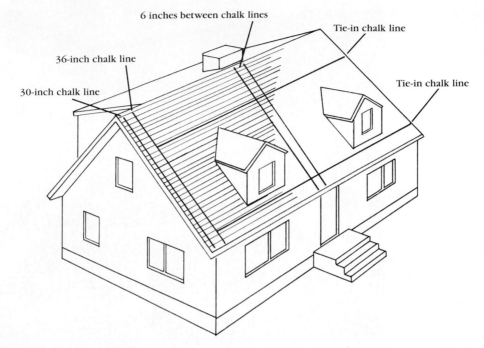

Fig. 6-53 Shingle up the rake and across the eaves for the straight pattern. Once you are past the bottom and the right side of the first dormer, snap chalk lines past the valley section of the dormer. Snap a horizontal tie-in chalk line, high-nail a horizontal tie-in shingle course, and fill in the pattern. Repeat the tie-in technique to shingle past the second dormer.

your mansard window flashing requirements with a knowledgeable materials supplier.

Begin shingling by applying a course of upside-down border shingles at the bottom edge of the mansard. Allow for the standard 1-inch overhang at the bottom of the facing board or trim board. Do not install drip edge or vertical border shingles along what would be the "rake" of a less steeply pitched, shingled roof surface.

Because you will be working from a ladder or a scaffold, use the straight pattern to install shingles on a mansard roof. The straight pattern allows you to nail the most shingles per run before you must move the ladder or scaffold. If practical, apply felt over the entire surface of the mansard, strike the 6-inch horizontal chalk line and the chalk line for the bottom border shingle, then reposition your ladder or scaffolding. Otherwise, you run the risk of misaligning the shingle cutouts as you snap a series of additional chalk lines up the surface of the mansard.

The professional roofers shown in Figs. 6-54A through 6-54J are using only the factory-made, 6-inch cutouts to align the shingle courses. I don't recommend this technique for do-it-yourselfers; too many things can go wrong.

Fig. 6-54A With a set of pump jacks and scaffolding to provide a stable platform from which to work, shingles can be installed on the deck of a mansard roof much like you would any other roof surface. Adjustable supports are secured with spikes at the eaves and fascia at both ends of the roof.

Fig. 6-54B Pump jacks are designed to be moved as the height of the work changes.

Fig. 6-54C   Each shingle at the window must be positioned, cut, and installed.

Fig. 6-54D   Apply felt with a hammer tacker, felt nails, or roofing nails.

Fig. 6-54E Trim the felt to fit tightly against the window.

Fig. 6-54F Because mansards are so visible, each shingle must be carefully aligned.

Fig. 6-54G Full shingles can be turned upside down, cut for position, trimmed, repositioned, and installed.

Fig. 6-54H The straight pattern is continued up the mansard roof.

Fig. 6-54I Another course is added to the straight pattern.

Fig. 6-54J The last two courses of the mansard shingles are trimmed at the tops of the shingles to fit where the mansard and the eaves meet. The first three courses on the top roof deck have been temporarily left out until the mansard has been shingled.

With the first course of felt and the first bottom border shingle in place, install at least one more course of felt up the mansard. Install successive courses of shingles until you reach the bottom of a mansard window or, if your mansard doesn't have a window, the eaves.

At a window, trim the tops and sides of successive courses just as you would for any of the obstacles described in chapter 7. As with a wall or a chimney, flashing around a mansard window must be watertight. Where the top course of mansard shingles meets the eaves shingles, trim the tops of the last course of mansard shingles so that last exposed course of shingles abuts the bottom of the eaves. See Fig. 6-54J.

Chapter **7**

Obstacles

*I*nstalling shingles around chimneys, walls, pipes, and roof vents requires careful planning and careful cutting and trimming of flashing and shingles. All of these obstacles are intrusions on roof surfaces; they must be watertight. Aluminum, galvanized, elastomeric, lead, and copper vent collars and *flashings* are commonly available at local building suppliers.

Specialty items such as chimney dressing, *cricket* flashing, wall *counterflashing*, or a complete chimney flashing kit can be obtained through the ABC Supply Company catalog. The counterflashing costs about $10 for four, 32-inch pieces and an aluminum chimney flashing kit—designed to fit 90 percent of all single-flue chimneys—costs about $30.

Chimneys

Old-Work Rake Chimneys

A chimney that is constructed along a rake must be flashed at the three sides where it meets the roof. Because the intrusion is at the rake, the shingle application pattern should be easy to continue. If you are reroofing over one layer of worn shingles, don't attempt to lift, remove, or replace the chimney flashing unless it is in poor shape.

For reroofing jobs, trim the shingles so that the shingles fit snugly against the flashing, avoid driving nails into the flashing, and then apply a generous amount of roofing cement or *mastic* around the chimney flashing. Use a generous amount of plastic-based roof cement at the back and at the side of the chimney where runoff will pass over the flashing.

New-Work Rake Chimneys

If you have torn off old shingles or if you are shingling new work and the chimney is at the rake, you must flash and counterflash the chimney. When you tear away shingles from a chimney, try to carefully bend up the counterflashing—without removing it—so that you can use it again (Fig. 7-1). If you must install new counterflashing, one common but not ideal method is to use 1-inch roofing nails to attach the counterflashing at the mortar joints.

Nail several runs of shingles until you reach the course just below the bottom of the chimney. As you add courses around the chimney, trim the tops and sides of the shingles to fit. Be sure to maintain the alignment of the cutouts. Trim the rake side of the shingles even with the border shingles. It is convenient to first position the shingle, cut the shingle to fit, then nail the trimmed piece of shingle in place.

As you nail shingles at the corners and sides of a newly constructed chimney, you must install flashing. Buy 5-×-7-inch aluminum step flashing or cut your own flashing. See Figs. 7-2 through 7-4.

Allow a 2-inch overlap with each piece of flashing and install the 7-inch portion of the step flashing against the roof surface. First position a shingle, nail a piece of step flashing in place (nail only on the 7-inch portion, using two nails at most), position the next shingle, then install another

Fig. 7-1 When you tear off shingles around a chimney, retain the flashing whenever possible.

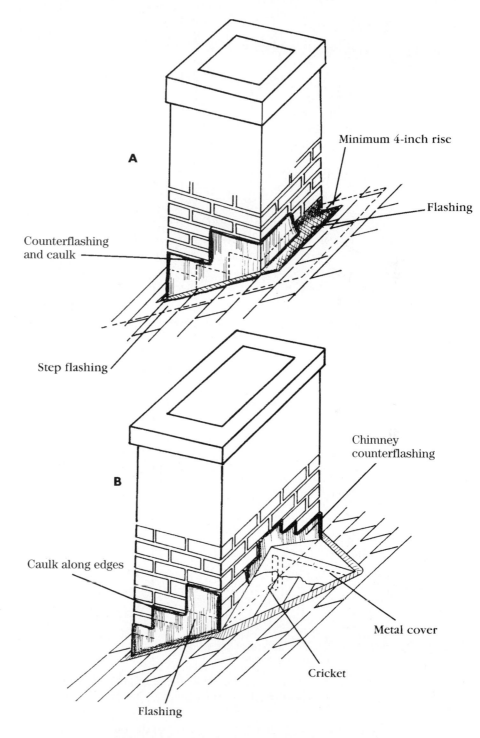

A

Minimum 4-inch rise

Flashing

Counterflashing
and caulk

Step flashing

B

Chimney
counterflashing

Caulk along edges

Metal cover

Cricket

Flashing

Fig. 7-2 Chimneys without (A) and with (B) a cricket.

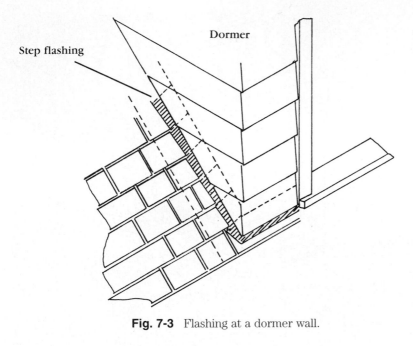

Fig. 7-3 Flashing at a dormer wall.

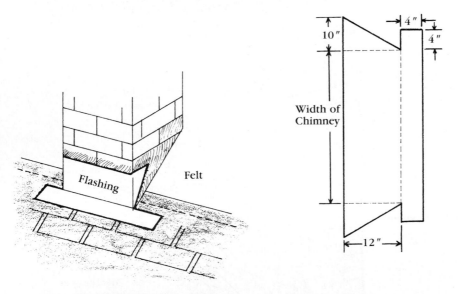

Fig. 7-4 Application of flashing at the front of a chimney.

piece of step flashing. Continue weaving the flashing and shingles. Remember that each piece of step flashing must overlap the one below it by 2 inches.

When you reach the shingle course that is even with the bottom of the chimney, nail the first shingle in place. Install a row of overlapped step

Fig. 7-5 With the corner of the chimney flashed, position the next course.

Fig. 7-6 Position flashing at the chimney corner and back.

flashing or one continuous piece of flashing flush against the bottom of the chimney.

At the bottom and top corners of the chimney, away from the rake, form watertight "boxes" of flashing. Use two pieces of step flashing to form a box. Use tin snips to cut one piece of flashing halfway into the crease. Do not detach the cut section. Bend the cut section so that when it is installed it will wrap around the corner of the chimney. Cut the second piece of flashing so that it will bend in the opposite direction. Install the box so that the higher piece is on top and water will flow over the flashing (see Figs. 7-5 and 7-6).

Install the next shingle flush with the side of the chimney. Maintain the pattern of shingle cutouts. As you install each piece of step flashing, make certain that the flashing is overlapped. The flashing should not extend past the bottoms of the shingles; in other words, there should be no metal showing except along the chimney side.

When you reach the top corner of the chimney, cut off the top section (all but the tabs) from one or two shingles, depending on the width of the chimney, and nail in place a strip that extends about 3 inches past the chimney corner. Trim the other side even with the rake border shingles.

Install flashing on top of the shingle strip along the back of the chimney, and nail the second corner flashing box in place. Trim the next few courses of shingles to fit snugly against the chimney and continue the application pattern.

Middle-of-the-Roof Chimneys

To successfully shingle around a chimney that intrudes upon the center of a roof section, plan the shingle pattern so that the courses meet past the top of the chimney. In addition, you will have to flash the chimney—this time on four sides—as described in the preceding section. Step flashing can be used around the chimney, or you can waterproof the chimney with flashing cut from galvanized valley material.

Read the section on dormers in chapter 6. Use the same basic techniques to tie in the shingles. Shingle the section directly below the chimney, strike chalk lines, and install a tie-in course (see Figs. 7-7 through 7-40).

The chimney must be watertight; so carefully install the flashing and avoid driving nails near the edges of the chimney. Apply plenty of roofing cement to all sides of the chimney and caulk the flashing seams. See Figs. 7-41 through 7-58.

Fig. 7-7 Tearing out worn chimney flashing with a pry bar.

Fig. 7-8 (Left) The old flashing is torn away from the back of the chimney.

Fig. 7-9 (Above) A claw hammer makes the work easier.

Fig. 7-10 (Left) Flashing can be cut from a roll of galvanized metal. Use a chalk line to mark an even line.

Fig. 7-11 (Above) Use tin snips to cut the metal along the chalk line.

Fig. 7-12 Bend the side flashing.

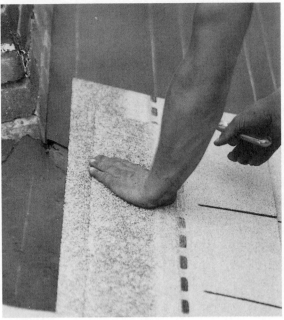

Fig. 7-13 Trim the shingle courses to fit tight against the bottom of the chimney.

Fig. 7-14 The shingle pattern is maintained.

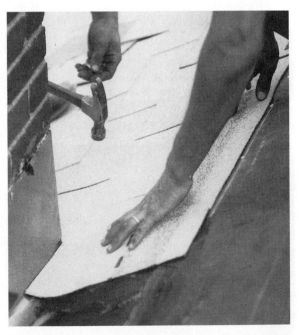

Fig. 7-15 Roofing cement provides additional protection at the back of a chimney.

Fig. 7-16   (Above left) Position the new flashing.

Fig. 7-17   (Above right) Use nails as a shortcut to securing counterflashing at the mortar joints. Brick masons recommend first opening the mortal joint, inserting the top of the counterflashing, and then remortaring the joint.

Fig. 7-18   (Left) Carefully position the flashing at the chimney corner.

Fig. 7-19   Position the side chimney flashing. The top of the flashing can be cut to give the appearance of steps.

Fig. 7-20   The flashing is placed on top of the shingle course at the chimney corner.

Fig. 7-21   At the chimney corner, one shingle is placed under the flashing, and the flashing is attached with roofing nails driven into the mortar joints. Masons recommend drilling out the old mortar, inserting the counterflashing, and remortaring the masonry joints.

Fig. 7-22 (Left) Trim the shingles so that they fit close to the flashing.

Fig. 7-23 (Below) Continue the shingle pattern.

Fig. 7-24   (Left) Cut on the back of the shingle so you can carefully size the tab.

Fig. 7-25   (Below) The tie-in course must be carefully planned.

Fig. 7-26. Shingle the courses until you reach the bottom of the middle-of-the-roof chimney.

Fig. 7-27 Mark a series of horizontal lines every 12 inches for the tie-in course.

Fig. 7-28 Snap chalk lines for the tie-in course.

Fig. 7-29 The straight pattern allows you to quickly shingle past the chimney.

Fig. 7-30 Once the shingle pattern is tied in, the area next to the chimney can be filled in.

Fig. 7-31 Flashing is installed on top of at least one course of shingles at the chimney base.

Fig. 7-32 Continue the shingle pattern at the chimney corner.

Fig. 7-33   Add a full shingle to the pattern.

Fig. 7-34 Weave flashing and shingles to waterproof the side of the chimney.

Fig. 7-35 Continue the pattern of weaving flashing and shingles.

Fig. 7-36   Add shingles to the pattern on both sides of the chimney.

Fig. 7-37   With the chimney corner flashing in place, position and cut a shingle strip to be placed behind the chimney.

Fig. 7-38   Install a strip of shingle at the back of the chimney for added waterproofing.

Fig. 7-39   Extend the shingle pattern past the chimney corner.

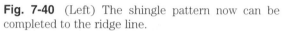

Fig. 7-40 (Left) The shingle pattern now can be completed to the ridge line.

Fig. 7-41 (Below) If the chimney flashing is not removed during reroofing, place a border shingle behind the chimney to provide flashing.

Fig. 7-42 Maintaining the 45 pattern past the chimney.

Fig. 7-43 Cut the tabs to fit snugly against the chimney side.

Fig. 7-44 (Left) Maintain the pattern with a cut shingle.

Fig. 7-45 (Below) Install a two-tab shingle so that another two-tab piece can be cut to fit against the side

Fig. 7-46 If the watermark is closer to the chimney than 6 inches, trim a two-tab piece to fit. Don't attempt to nail a very small piece of shingle next to the chimney.

Fig. 7-47 Add courses to maintain the 45 pattern along the side of the chimney.

Fig. 7-48 Install tabs to maintain the pattern.

Fig. 7-49 Trim the corner shingle to fit.

Fig. 7-50 The shingle pattern is continued past the chimney.

Fig. 7-51　Position the tab before nailing.

Fig. 7-52　Cut the tab to fit the corner.

Fig. 7-53　Install the corner shingle and continue the pattern.

Fig. 7-54　Fill in the courses by lifting tabs.

Fig. 7-55 Courses above the tie-in shingles have been high-nailed.

Fig. 7-56 Apply roofing cement along the chimney sides.

Fig. 7-57 Apply plenty of roofing cement at the back of the chimney.

Fig. 7-58 Apply caulk to flashing seams.

Vent Flanges and Pipe Collars

Aviation snips or a utility knife with a hook blade can be used to cut shingles to fit around vents, stacks, and pipes. Other obstacles, such as a roof-mounted evaporative cooling unit or air conditioner, can be shingled without chalk lines or elaborate tie-in courses.

Use a small, hand-operated pump jack and several planks to raise and brace a roof-mounted evaporative cooler several inches above the roof deck during tear-off work. While the area under the cooling unit is being reshingled, raise the support brackets only a few inches to allow the courses to be positioned under the brackets that normally support the unit. Once the shingle courses are installed beneath the cooling unit, remove the pump jack and planks and set the brackets so that they rest on top of the shingles. See Figs. 7-59 and 7-60.

A vent or pipe, sometimes called a jack, must have a *vent collar* or flange that diverts water away from the opening in the roof deck that accommodates the ventilation or plumbing pipe. On many vents, the flange is part of the unit that fits over, surrounds, or otherwise seals the opening in the roof deck. Pipes and vents are commonly flashed with a metal or polyurethane collar; the flange usually is part of the collar (Fig. 7-61). Shingle the roof section until you have at least one full course below the obstacle. See Figs. 7-62 and 7-63.

➡ **Application Tip** To avoid the two most common errors when installing shingles around vents and pipes, face the tapered neck of the pipe collar so that it points down the roof when installed. Second, install at least one course of shingles directly below the pipe under the flange. It might be necessary to place a portion of more than one course under the flange. A substantial leak will result if a flange is improperly installed.

Trim the circular cuts in the one-tab shingles around the pipe. Where possible, temporarily push the shingle tab into position, make scratch marks at the appropriate cut locations, and then use a utility knife or tin snips to make partial cuts. Place the tab in position and determine if more needs to be cut. The shingles should fit as close to the pipe and collar as possible (Figs. 7-64 through 7-68). Apply a generous amount of roofing cement where the collar and the shingles meet and over any exposed nail heads.

Fig. 7-59 A hand-pumped jack and planking temporarily support an evaporative cooler above the roof deck while debris is removed and the roof is reshingled.

Fig. 7-60 Once the deck beneath the roof-mounted cooling unit has been reshingled, the support brackets are placed on top of the shingle courses.

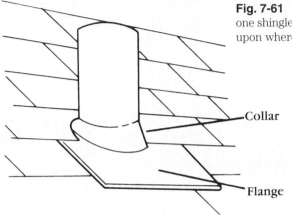

Fig. 7-61 A flange must be installed on top of at least one shingle. The amount of exposed flange will depend upon where the shingle course intersects the pipe.

Collar

Flange

Fig. 7-62 An old flange can be used with new shingles.

Fig. 7-63 Install the flange over the shingle course.

Fig. 7-64   Position and mark the shingle for cutting.

Fig. 7-65   Trim the shingle around the pipe flange.

Fig. 7-66 Trim the next shingle.

Fig. 7-67 Shingles should fit around, but not over, the rounded portion of the flange. Many roofers trim the shingles, as shown, snug against the flange and then apply a generous layer of roofing cement around the base. Others trim an opening in the shingles at the base of the flange to prevent potential buildup of ice or runoff.

Fig. 7-68 Continue the pattern past the pipe.

Walls

If you are shingling over one layer of worn shingles, do not lift or remove old step flashing from alongside walls unless the flashing is very worn. As you install the shingles, trim the shingles as close to the wall as possible.

On new work, install 5-x-7-inch step flashing with the 7-inch portions flat against the roof (see Figs. 7-69 and 7-70). At the bottom corner where the wall and the roof meet, install box flashing as you would at the corner

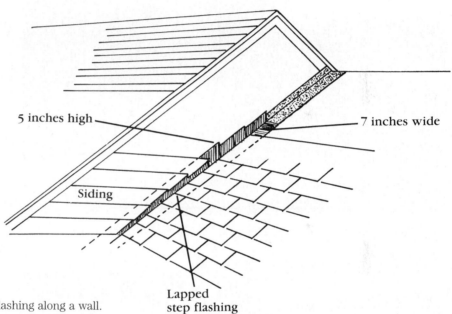

5 inches high

7 inches wide

Siding

Lapped
step flashing

Fig. 7-69 Install step flashing along a wall.

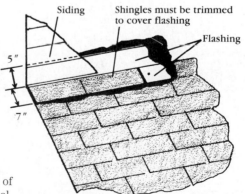

Siding

Shingles must be trimmed
to cover flashing

Flashing

5"

7"

Fig. 7-70 Install step flashing or a continuous strip of metal against a vertical wall to ensure a watertight seal.

of a chimney. Weave the shingles and step flashing as you make the run. See Figs. 7-71 through 7-78.

At the ridge line, use tin snips to cut a piece of step flashing halfway along the crease. Cut another piece of step flashing along the opposite side of the crease. Fold the edges down, slip the pieces together, and form a

Fig. 7-71   Install a border shingle where the rake and wall meet.

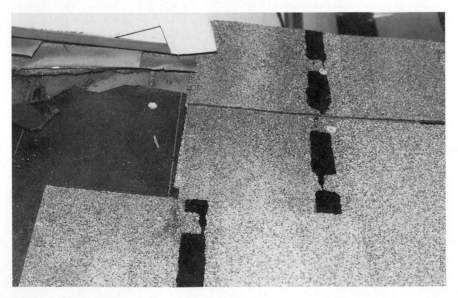

Fig. 7-72   Continue the shingle pattern by installing tabs.

flashing "tent" at the ridge line. Nail down the tent after you have shingled and flashed both sides of the roof section.

Where shingle courses intersect a vertical wall, such as beneath a dormer, install at least two courses of shingles under the bottom edge of the flashing that protects the wall. The wall flashing can be overlapped step flashing or a continuous strip of metal, must provide 5-x-7-inch coverage, and must be counterflashed behind the wall's siding. See Figs. 7-79 through 7-87.

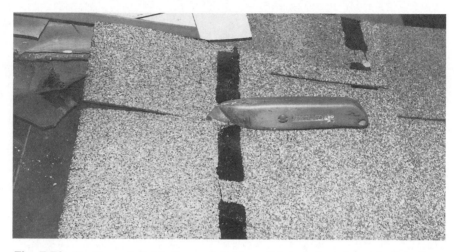

Fig. 7-73   Position a full shingle, mark it where it is to be cut, then trim it to fit.

Fig. 7-74   Install the trimmed shingle (Fig. 7-73) under the first piece of step flashing. On this reroofed section, the step flashing was not completely removed from along the wall. Nails holding the flashing in place were removed so that the shingle courses could be woven with the flashing.

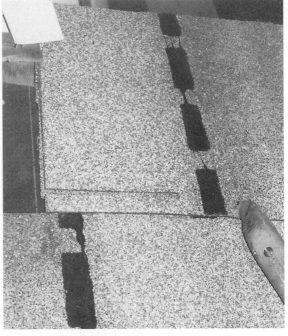

Fig. 7-75 By turning the shingle upside down, you can easily find the proper place to cut.

Fig. 7-76 Again, turn the shingle upside down to mark the remaining portion of shingle to be trimmed.

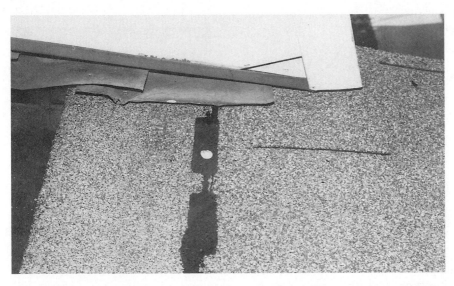

Fig. 7-77 With the shingle properly trimmed to fit against the wall, place the second piece of step flashing over the shingle and nail about $1/4$ inch from the edge. Don't nail the step flashing too close to the wall. Compare the position of the flashing with that shown in Figs. 7-74 and 7-78.

Fig. 7-78. Continue weaving step flashing and shingles. The flashing should be even with or slightly beyond the watermarks.

Fig. 7-79 The shingle pattern approaches the left side of the dormer valley and wall.

Fig. 7-80   At least two full shingles must be placed under the rake of the dormer. Note the minor damage to the roof deck.

Fig. 7-81   Minor repairs were made to the roof deck and the shingle pattern extended beneath the dormer rake.

Fig. 7-82 Install full shingles as you approach the wall.

Fig. 7-83 Fit the second-to-last shingle course under the lip of the flashing with the metal covering the top few inches of the full shingle. Also, install the top portion of the metal flush against the wall and counterflashed by the siding.

Fig. 7-84   The shingle pattern approaches the right side of the dormer valley and wall.

Fig. 7-85   Position a full shingle under the rake.

Fig. 7-86 Position the second-to-last course under the flashing and under the dormer edge.

Fig. 7-87 Trim the top course to fit under the flashing and against the wall.

Chapter **8**

Low-Slope Roofing

*R*oll roofing and foam roofing products are appropriate choices for low-slope roofs with as gentle a pitch as 1 inch per foot, such as sheds, porches, deck covers, and garages, and are best applied to surfaces where appearance is not apparent or important because it is not aesthetically appealing. *Roll roofing* can be applied horizontally or vertically, but vertical installation is not as watertight because the long seams are subject to seepage. The advantages of using roll roofing include low cost and ease of application. Foam roofing products must be applied by trained, approved contractors.

One roll-roofing product, E-Z Roof, made by Tramac Roofing Systems, has the added advantage of a peel-and-stick, self-adhesive backing that provides a quick bond without the need for spreading gallons of roofing cement. Similar ice-and-water-shield membranes can be used to waterproof eaves and valleys.

Torch roll roofing is an increasingly popular product applied by professional roofers with the aid of a propane torch. Heat from the torch activates a sealant built into each roll of the roofing material. See Figs. 8-1 through 8-16. The underlayment for torch-sealed roll roofing is often specified to be heavier than 15-pound felt, but the same techniques described in chapter 4 are used to prepare the roof deck.

Roll roofing is manufactured with the same basic materials as asphalt shingles but its weight per 100 square feet is 90 to 180 pounds. Roll roofing comes in 36-foot-long, 3-foot-wide strips—and it will wear out in about half the time as most shingles. Asphalt, fiberglass-based shingles range in weight from 210 to 325 pounds per 100 square feet. The difference in weight is the main reason for roll roofing's short life span.

Fig. 8-1   Cover a flat-roof deck with an underlayment and install drip edge.

Fig. 8-2   Temporarily roll out the first course of roll roofing and trim with a utility knife.

Fig. 8-3 Adjust the first course to remove any wrinkles and to confirm the proper length.

Fig. 8-4 Reroll the material toward the center of the roof deck.

Fig. 8-5 As the first course is slowly rolled into place, heat from a propane torch is used to activate the built-in sealing compound.

Fig. 8-6 At the rake and eaves, heat the membrane to form a seal even with the drip edge.

Fig. 8-7 Reroll the second half of the starter course. Next, apply the remainder of the first course working from the center of the roof toward the opposite rake.

Fig. 8-8 Position the second course so that it overlaps the starter course by several inches.

Fig. 8-9 Pull the second course tight to remove any wrinkles in the material.

Fig. 8-10 Use a utility knife to trim the second course at the rake.

Fig. 8-11 Trim the second course and temporarily place it in position.

Fig. 8-12 Reroll the second course toward the center of the roof deck.

Fig. 8-13   Heat from the propane torch activates the sealant for the second course.

Fig. 8-14   Where sheets of roll roofing do not reach the full length of the deck, do not match up vertical overlaps of successive courses.

Fig. 8-15 Allow for at least 6 inches of overlap where courses meet. The propane-fueled flame indicates why manufacturers prefer that only authorized applicators install torched roll roofing.

Fig. 8-16 A gloved hand holding a trowel positions the last section to be torched.

To prepare a roof surface for roll roofing, follow the guidelines as described in the chapter 4 sections on tearing off worn materials, laying felt, and installing drip edge. You will need the following tools and equipment:

- stiff-bristle broom
- ladder
- nail bar
- nail pouch
- roofing cement
- roofer's hatchet
- roofing nails (1-inch galvanized or aluminum)
- tape measure
- trowel
- utility knife or aviation snips

Single-Coverage Roll Roofing

First Course

With the felt and drip edge in place, apply a horizontal starter course of roll roofing the entire length of the eaves. If the starter course is longer than 36 feet, overlap the joints by 8 inches. The starter course should extend over the rakes and the eaves by 1 inch (see Fig. 8-17).

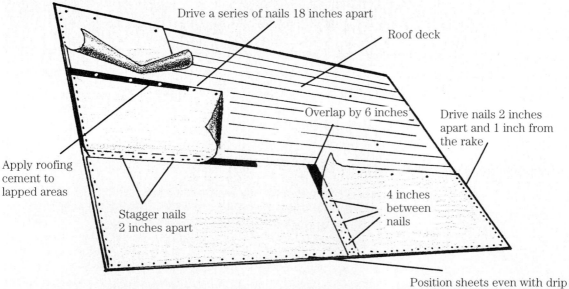

Fig. 8-17   The basics of applying roll roofing.

Along the top of the starter strip, about $1/_2$ inch from the top, nail the course every $1^1/_2$ feet. At the bottom and at the rakes, drive nails 2 inches apart and about 1 inch from the edges.

Second Course

Position the second course so that it overlaps the top of the starter course by at least 2 inches. Most roll roofing manufacturers include guidelines on their products for easy alignment of courses.

If the roof surface is longer than 36 feet, do not align the joints in the courses or leaks will result. Where joints are necessary, overlap the joints by about 1 foot. Use a trowel or a stiff-bristle brush to apply a smooth coat of roofing cement, then embed the overlapping sheets. Nail the lap joint at 2-inch intervals.

Third to Final Courses

Continue applying courses until you reach the ridge or a wall. If the roof has a ridge, trim the top course even with the ridge on one side. Lap the last course from the other side of the roof and nail it snugly against the deck.

Using a chalk line and tin snips or a utility knife, make ridge capping by cutting a strip of roll roofing in half, lengthwise. When you position the capping course, make certain the capping strip is straight and without bulges or wrinkles.

If the roll-roofing courses must abut a wall, cut L-shaped galvanized step flashing, about 42 inches long, for each course. Weave the roofing and the metal so that the flashing method resembles the techniques used to waterproof the sides of a chimney, as described in chapter 7.

Double-Coverage Roll Roofing

To allow for double coverage with roll roofing, manufacturers produce *selvage-edge roll roofing*, which provides 17 inches of mineral-surface roofing and 19 inches of saturated felt in one 36-inch-wide, 36-foot-long sheet. Some selvage-edge manufacturers require the use of cold, asphalt-based roofing cement with their products; other manufacturers require the use of hot asphalt. Cold asphalt-based roofing cement makes the job practical for the do-it-yourselfer.

Horizontally applied selvage-edge roll roofing can be used on a roof deck with a rise of 1 inch or more per foot. Prepare the roof surface as described in the chapter 4 sections on tearing off worn materials, laying felt, and installing drip edge. In addition to the tools and equipment listed at the beginning of this chapter, you'll need an additional small stiff-bristled broom for applying the type and quantity of roofing cement recommended by the selvage-edge manufacturer.

First Course

With the felt and drip edge in place, make a horizontal starter course by cutting the selvage portion of a strip long enough to reach the entire length of

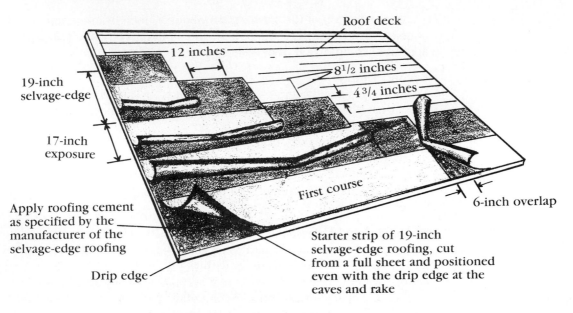

Fig. 8-18   Installing double-coverage roll roofing.

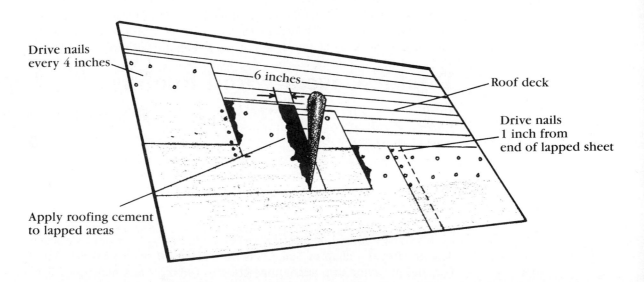

Fig. 8-19   Applying roofing cement to the lapped areas.

the eaves. Save the mineral-surface portion of the roll for the last course. If the starter course must be longer than 36 feet, overlap the joints by about 8 inches. The starter course should overlap the rakes and the eaves by 1 inch.

Secure the selvage starter course by driving roofing nails at 6-inch intervals along the rake, along the eaves, and staggered across the center.

To complete the first course, roll out a 36-inch-wide strip of selvage-edge roll roofing long enough to cover the starter course. Align the material so that the mineral-surface (granule) portion of the strip exactly matches the coverage of the starter strip. Nail only the selvage portion in place. Do not drive any nails in the mineral-surface portion of the roll roofing.

When the course is in position, very carefully lift up the mineral-surface portion and gently flip it over to expose the selvage starter course. Use a short-bristle broom to apply roofing cement, as specified by the manufacturer, to the selvage starter course. Gently flip the mineral-surface portion of the roll roofing back into position. Be careful not to tear the strip. See Figs. 8-18 and 8-19.

Second Course

Apply the next course of shingles and succeeding courses by lapping the previous selvage-edge layer with the mineral surface of the next strip of roll roofing. If joints are necessary, overlap the edges by about 8 inches and apply roofing cement along the joint. Never align joints in successive courses or leaks will develop.

Last Course

When you are ready to apply the last course of mineral-surface material, retrieve and nail in place the strip left over from the selvage-edge starter course. After all the courses have been applied, use your feet to ensure that all the sheets of roll roofing have good adhesion.

Polyurethane Foam

For problem-prone, low-slope roofs and flat roofs over decks, porches, and carports, *polyurethane foam roofing* can be sprayed on by professional, factory-trained applicators to provide a seamless, lightweight, waterproof surface. The chemicals that form the wet foam must be mixed at the job site using special equipment.

Before the layers of foam are applied, the roof deck must be cleared of dust, dirt, leaves, twigs, nails, and other debris. A starter-course layer of foam is first applied to the deck, rough edges are filled in, the edges are smoothed with a power sander, more polyurethane insulation is added, the system is allowed to dry for several days, and then the roof is sealed with another layer of sprayed-on elastomeric membrane as a top coat.

The first two layers go on quickly and the first portion of a typical job can be completed in a few hours. See Figs. 8-20 through 8-33. After a few days, when the initial layers have hardened, the third layer also can be sprayed on quickly. A contractor will most likely have to return every five years or so to touch up the sealer.

Fig. 8-20 A portable leaf blower is used to quickly clear the deck of leaves and other debris.

Fig. 8-21 Drip edge with a "gravel stop" lip is installed along the eaves of the porch deck to ensure a uniform depth for the foam and an even appearance at the deck edges.

Fig. 8-22 The applicator must wear protective clothing and a mask to guard against misdirected or wind-blown chemical spray. Adjust and test the mixture texture before applying it to the roof surface.

Fig. 8-23   A light spray from the nozzle quickly transforms into a coating of water-proof foam.

Fig. 8-24   Directing the foam into otherwise hard-to-reach corners.

Fig. 8-25 (Above left) A "dead" valley—created when a new section was added to the original home— is difficult to waterproof with conventional roofing materials. Foam applied directly to the planking makes the job easy.

Fig. 8-26 (Above right) The valley's second coating is applied directly over the base foam.

Fig. 8-27 (Left) Two courses of planks temporarily have been removed from the main roof deck to provide access in order to seal the back of the porch

Fig. 8-28 Once obstacles are dealt with, the main porch deck can be foamed very quickly.

Fig. 8-29 A paper barrier is temporarily positioned at the eaves to block the overspray from leaving the roof deck.

Fig. 8-30 (Left) A power sander is used to smooth even the foam at the deck edges.

Fig. 8-31 (Below) At the eaves, the edges are filled in with the help of a paint applicator, and the second coating for the porch deck is started.

Fig. 8-32   The second coating can be applied in only a few minutes.

Fig. 8-33   The edge of the deck receives a careful coating as finishing touches are applied.

Chapter **9**

Wood Shingles
and Shakes

Before you decide to use wood shingles or shakes on your home, check that your local building code permits them. In California, some local building codes ban reroofing with wood—that is, replacing a worn wood roof with a new wood roof—because of concerns about potential fires. In some California locations, the hazards from brush fires and the resulting flying embers are so prevalent that insurance companies are not only unwilling to insure homes with wood roofs, they won't insure homes adjacent to residences with wood roofs—even when such adjacent homes have nonwood roofs.

In a direct response to the 1993 firestorm that destroyed some 400 Laguna Beach, California homes, Boulder, Colorado officials adopted an ordinance that bans all types of wood roofing within city limits. Boulder city officials felt the restrictions were a reasonable precaution against wildfires for vulnerable residential areas.

Where climates are suitable, redwood or cedar wood shingles or hand-split shakes do make attractive alternatives to asphalt fiberglass-based shingles. Compared to typical asphalt fiberglass-based shingles, however, the costs of wood shingles or shakes is higher. For #1 Blue Label wood shingles that are cut from 100 percent clear, edge-grain heartwood, you can expect to pay about twice the cost of the heaviest asphalt-fiberglass shingles. On the other hand, cedar shingles have about twice the insulating value as asphalt shingles.

Wood shingles are produced in uniform 16-, 18-, and 24-inch widths. Wood shakes are split to expose at least one natural-grain textured surface, and therefore, present an irregular pattern when installed. Packed in bundles (Fig. 9-1) that are easy to handle, wood shingles and shakes are applied using the basic principle of roofing: overlapped layers of water-shedding materials.

Fig. 9-1 Individual bundles of wood shingles are easy to handle. The large volume of materials required to roof a typical home makes careful planning essential.

Installation techniques for shakes and shingles are similar. Exceptions are that an 18-inch-wide *interlay* strip of 15-pound felt must be applied beneath each course of shakes (interlayed felt is optional for wood shingles), and the amount of exposure to the weather for shakes is greater than for shingles. While shakes or shingles can be applied to a deck without plywood (spaced sheathing) or standard sheathing, or over one layer of worn asphalt shingles or roll roofing, wood shingles should be installed over furring strips or fiberglass mesh to provide the proper ventilation that will ensure long shingle life. I don't recommend using an underlayment of felt instead of the interlaced method in humid climates; condensation under the shakes will shorten the life span of the roof.

Tools and Equipment

Using the proper tools and equipment for the job will make shingling work easier and faster. Along with the items described in chapter 2, you should use a 4-foot-long straightedge tacking board (Fig. 9-2) to align the courses as you work across a roof section. A wood roof seat is useful on a roof with a not-too-steep pitch, and a hand-held rechargeable power saw is convenient for angling and trimming shingles or shakes.

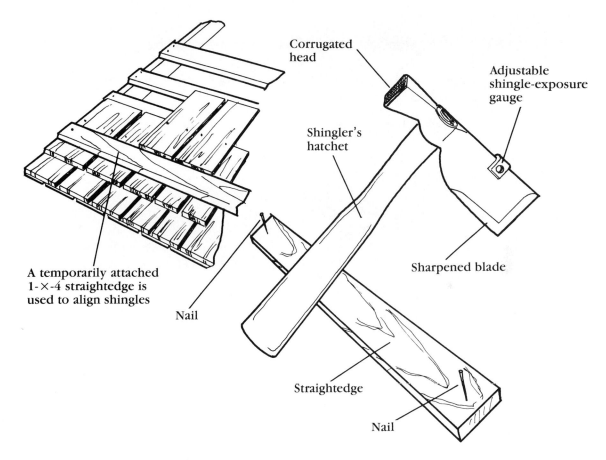

Fig. 9-2 A straightedge tacking board and shingler's hatchet.

Hatchet Gauge

The adjustable *gauge* on a shingling hatchet or a similar gauge on a pneumatic nailer provides a convenient way to check shingle exposure. The gauge can be set, for example, at $7\frac{1}{2}$ inches from the head of the hatchet (Fig. 9-3). When a shingle is in position (Fig. 9-4), you can use the gauge to confirm when a shingle is ready to be nailed in place (Figs. 9-5 and 9-6).

Roofer's Seat

A roofer's seat (Fig. 9-7) can only be used to install wood shingles or shakes. The seat would mar fiberglass shingles and it cannot be used on a metal roof. You can make your own seat by cutting an 18-inch-long base (seat) and a 15-inch-long support base from 1-×-12 lumber. You will also need strips of wood for the "gripper" nails.

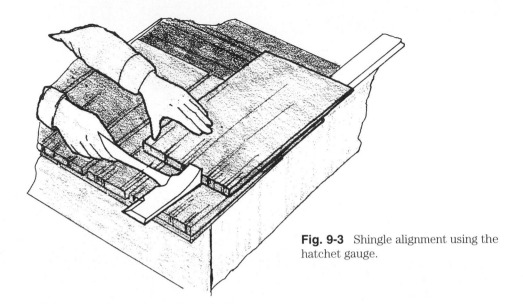

Fig. 9-3 Shingle alignment using the hatchet gauge.

Fig. 9-4 Position the first shingle in a new course so that the top several inches of the shingle slides under one layer of felt. About half of the shingle goes on top of the next layer of felt, and the remainder of the shingle laps the underlying shingle course and is exposed to the weather.

Fig. 9-5 Proper shingle spacing.

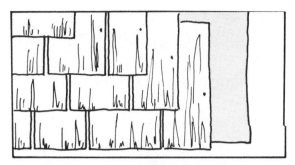

Fig. 9-6 Allow a $1/_4$-inch expansion gap between shingles and avoid alignment of vertical joints by at least $1^1/2$ inches.

1. Place the base of the seat on the roof facing the ridge. With the seat side on top, align the front of the seat and the support base.
2. Lift the back of the seat so that it is level, and then measure the length of boards needed to fit the sides.
3. Cut and install the pieces between the seat and base.
4. Cut several 3-inch-wide, 10-inch-long strips of plywood and drive several 1-inch roofing nails through each strip.
5. Nail the strips to the base of the slat with the nail points facing out. The nails will grip the roof and keep the seat in position as you work.

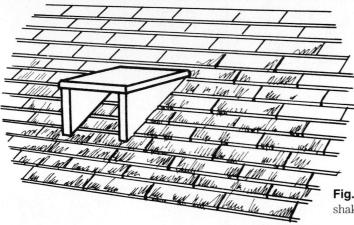

Fig. 9-7 A roof seat helps you to install wood shakes or shingles.

Nails

Each shingle or shake should be secured with two and only two nails. The length of the galvanized nails used varies with the size of the roofing material and the type of base they will cover. For 16- and 18-inch shingles, use 3d nails, and for 24-inch shingles, use 4d nails. Use 6d nails for shakes unless the shakes are applied over one layer of worn roofing; then use 7d or 8d nails. At hips and ridges, always use 8d nails. Table 3-2 lists nail sizes.

Stains and Coatings

Cedar wood shingles and shakes can be left uncoated to turn a natural attractive gray or you can apply one of a variety of clear coatings or stains. If the shingles or shakes are left untreated, they can sometimes develop an unattractive or uneven appearance.

A clear coating provides some protection against wood rot and helps repel moisture. A semitransparent stain allows the wood texture and grain to show and provides a more Colonial look. So that paint can be applied, which would ordinarily seal the wood and defeat water-shedding properties, untreated cedar needs one or more coats of bleed-resistant primer. Follow the manufacturer's instructions according to label directions.

Fungicides intended to inhibit mildew and moss should be applied carefully following the manufacturer's precise instructions for the product. An alternative is to buy treated shingles or shakes.

Estimating Materials

Determining the amount of materials you'll need to roof your home with wood shingles or shakes is more complicated than asphalt-fiberglass shingles. Because the amount of wood shingle or wood shake exposure varies with the pitch of your roof, you must first determine the roof's total

Table 9-1 Recommended Exposure for Wood Shingles and Shakes.

Do not use wood shingles on a roof with less than 3" in 12" pitch			
Shingle length	*Shingle thickness*	*Slope: 3 in 12*	*Slope: 4 in 1*
16"	5 butts in 2 inches	$3^3/_4$" exposure	5" exposure
18"	5 butts in $2^1/_4$ inches	$4^1/_4$" exposure	$5^1/_2$" exposure
24"	4 butts in 2 inches	$5^3/_4$" exposure	$7^1/_2$" exposure
Shake length			
18"		not recommended	$7^1/_2$" exposure
24"		not recommended	10" exposure
Length	*Exposure*	*Coverage*	
16" Shingle	$7^1/_2$"	150 square feet	
18" Shingle	$8^1/_2$"	154 square feet	
24" Shingle	$11^1/_2$"	153 square feet	
18" Shake	$8^1/_2$"	85 square feet	
24" Shake	$11^1/_2$"	115 square feet	

square footage, and then add a percentage to compensate for the pitch (Table 9-1).

Include in your estimate total one square of shingles for every 100 linear feet of hips and valleys. For starter shingles, one square will cover about 240 linear feet. Allow two squares of shakes for every 120 linear feet of hips, valleys, and starter course. It takes four bundles to make a square (100 square feet) of wood shingles. There are five bundles per square of shakes.

Shingle Application

To apply wood shingles over a layer of worn wood or asphalt shingles, first cut back about 6 inches of the old roofing along the eaves and rakes to provide a neat appearance. Nail strips of lumber along the rakes and eaves so that all of the roof-section surfaces are level. If you have decided to use drip edge, install it now along the eaves and rakes, as described in chapter 4.

Ventilation Strips

If the shingle exposure is $5^1/_2$ inches or more, use 1-×-3s for ventilation furring strips; use 1-×-2s if the shingle exposure is less than $5^1/_2$ inches. Nail the furring strips parallel with the eaves and spaced at equal distances so that the shingle exposure remains the same for every course.

Starter Course

At both rakes of the roof section, position a shingle so that the butt over-hangs the eaves by $1\frac{1}{2}$ inches. The rakes should have a 1-inch overhang. To secure the first end shingle, drive nails about $1\frac{1}{2}$ inches above the expo-sure line. Install the starter course across the eaves (Figs. 9-8 and 9-12) by spacing the shingles $\frac{1}{4}$ inch apart. The joints (Fig. 9-9) permit the wood to expand during hot weather, which prevents the wood from buckling. The starter course must be doubled, and it can be tripled if you prefer a more textured appearance at the eaves. Offset all vertical joints between the shingles by $1\frac{1}{2}$ inches.

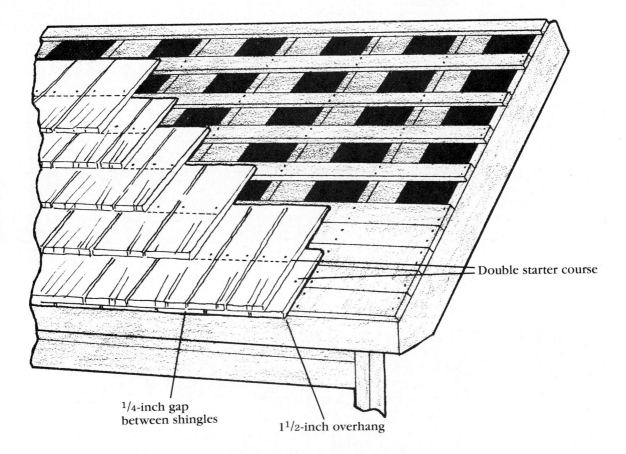

Double starter course

1/4-inch gap
between shingles

1 1/2-inch overhang

Fig. 9-8   Ventilation strips installed with wood shingles.

Fig. 9-9 Shingle spacing and alignment of courses.

Shingles are spaced 1/4 inch apart, providing room for expansion during hot weather

Do not align joints on successive courses

Shingle lap must be at least 1½ inches on successive courses

Fig. 9-10 For subsequent courses, the joints for no three adjacent courses should align.

Subsequent Courses

With each subsequent course (Fig. 9-10), no joints in any three adjacent courses should align. For hips (Fig. 9-11) and ridges, use special factory-assembled units. Saw shingles that extend into valleys so that the shingles

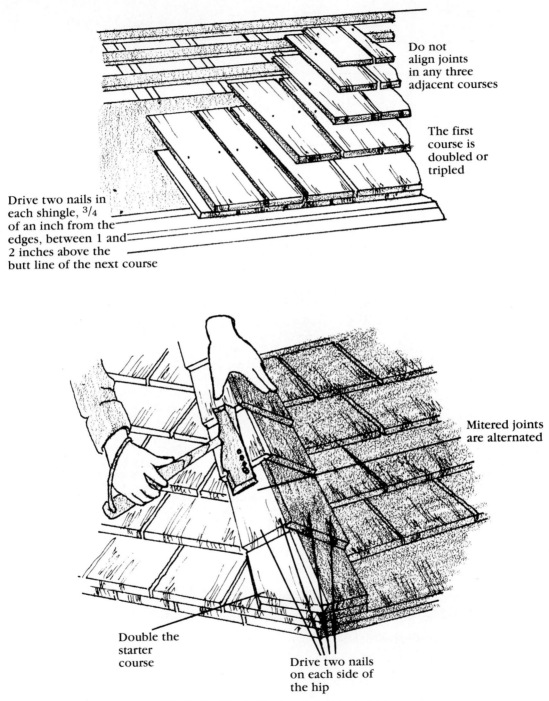

Do not align joints in any three adjacent courses

The first course is doubled or tripled

Drive two nails in each shingle, 3/4 of an inch from the edges, between 1 and 2 inches above the butt line of the next course

Mitered joints are alternated

Double the starter course

Drive two nails on each side of the hip

Fig. 9-11   Wood shingle application techniques.

Width of valley from centerline
varies with the pitch of the roof:
7 inches to 10 inches

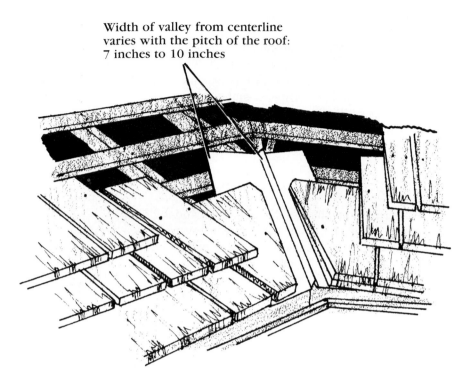

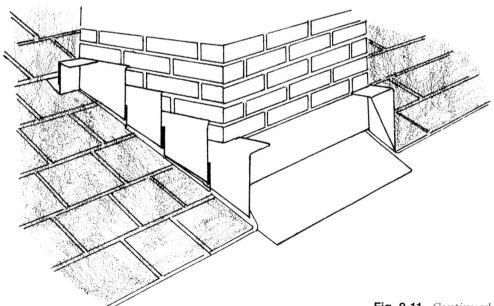

Fig. 9-11 *Continued.*

fit the angle of the valley center. Do not install shingles parallel with the angle of a valley.

Valley flashing should extend 10 or more inches on either side of the center of the valley. Walls, vents, and chimneys are potential trouble spots. See chapter 7 for detailed information on how to deal with these common roof obstacles. Be especially careful when you install flashing and wood around pipes. Remember that the top of the flange must be covered by the roofing material and the bottom of the flange must go over the top of one or two courses of the roofing material (Figs. 9-12 and 9-13).

Shake Application

To apply wood shakes over a layer of worn shingles, first cut back about 6 inches of the old roofing material along the eaves and rakes to ensure a neat appearance. Nail strips of lumber along the rakes and eaves so that all of the roof-section surfaces are level. If you use drip edge, install it now along the eaves and rakes, as described in chapter 4.

Starter Course

Roll out and install a 36-inch-wide course of felt the length of the eaves and trimmed at the rakes. Either 15- or 18-inch shakes can be used for the underlying starter course.

At both rakes of the roof section, position a shingle with the butt overhanging the eaves by 2 inches. The rakes should have a $1^1/_2$-inch overhang. To secure the first end shingles, drive nails about $1^1/_2$ inches above the exposure line. Install the starter course across the eaves by spacing the shingles $1/_2$ inch apart. The starter course must be doubled, and it can be tripled if you prefer a more textured appearance at the eaves. See Figs. 9-14 and 9-15.

Offset the vertical joints between the shakes by at least $1^1/_2$ inches; leaks will result from the improper alignment of joints. Also, remember that the joints permit wood to expand during hot weather, which prevents the wood from buckling.

Subsequent Courses

Cut an 18-inch strip of felt the length of the starter course and install it across the top of the shakes. The bottom of the 18-inch felt course should lap the shakes 20 inches above the butt line of the shakes. Use a straight-edge tacking board or the gauge on your roofing hatchet to ensure the proper shake exposure before you install a course of shakes over the felt.

You can save time by laying three or four more courses of 18-inch-wide felt strips on the roof section, then tucking the shakes under the felt as you apply each course. Don't apply felt to an area larger than what you'll be able to work on in a day. Use 1-inch roofing nails to secure only the top edge of the felt. If you nail the felt at any other points, you will not be able to install

Fig. 9-12 Roof vents and turbines must be flashed with flanges over the top of at least one course of shingles. Use a hand-held, rechargeable power saw or a circular saw to trim shingles to fit.

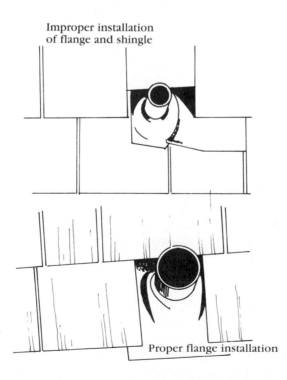

Improper installation of flange and shingle

Proper flange installation

Fig. 9-13 Install vent flanges under at least one course.

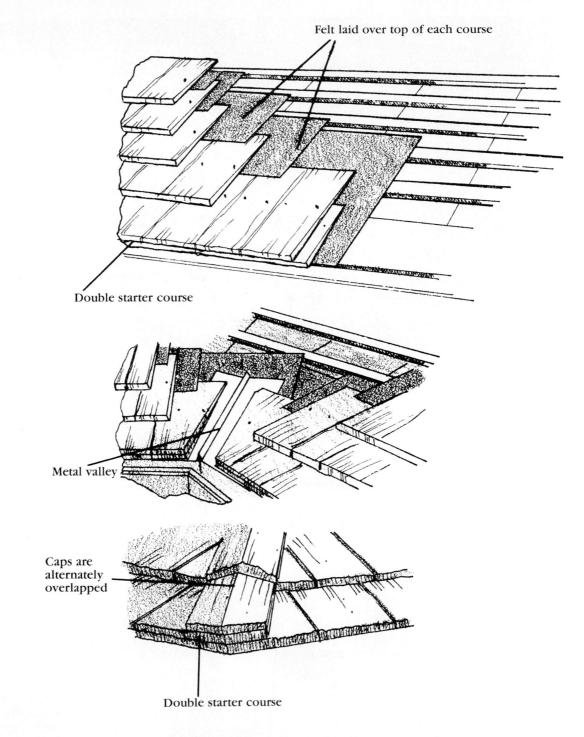

Felt laid over top of each course

Double starter course

Metal valley

Caps are
alternately
overlapped

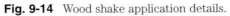

Double starter course

Fig. 9-14 Wood shake application details.

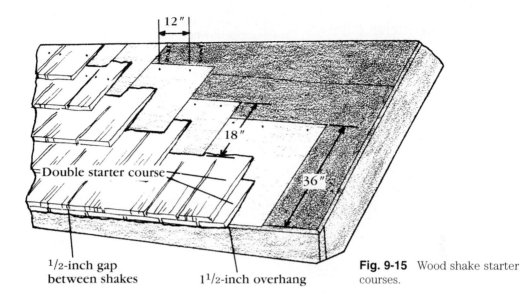

12″

18″

36″

Double starter course

½-inch gap
between shakes

1½-inch overhang

Fig. 9-15 Wood shake starter courses.

the shakes properly. Be extremely careful when you apply the felt strips. Never walk on felt that has not been securely nailed.

Continue interlaying felt with shake courses as you roof the building. No joints in any three adjacent courses should align. For hips and ridges, use special factory-assembled units. Saw shingles that extend into valleys so that the shingles fit the angle of the valley center. Do not install shingles parallel with the angle of a valley.

Panelized Roofing

Installing individual shakes or shingles can take three to five times as long as installing asphalt fiberglass-based shingles. Two alternative materials are available, however, that can speed up installation.

Shakertown's cedar shingles and shakes are bonded with waterproof adhesives to form three-ply, 8-foot-long panels. The panels do not need sheathing and can be nailed directly to studs. Shakertown two-ply panels can be applied to sheathing or to furring strips. A self-aligning feature and the need for only two nails per panel per rafter speeds the work, and no straightedge or gauges are needed to keep the courses aligned. See Fig. 9-16. Complete application instructions are included with each bundle of panels.

Another product that speeds installation is Masonite's fiberboard-panel roofing product, Woodruf, which is manufactured from highly compressed

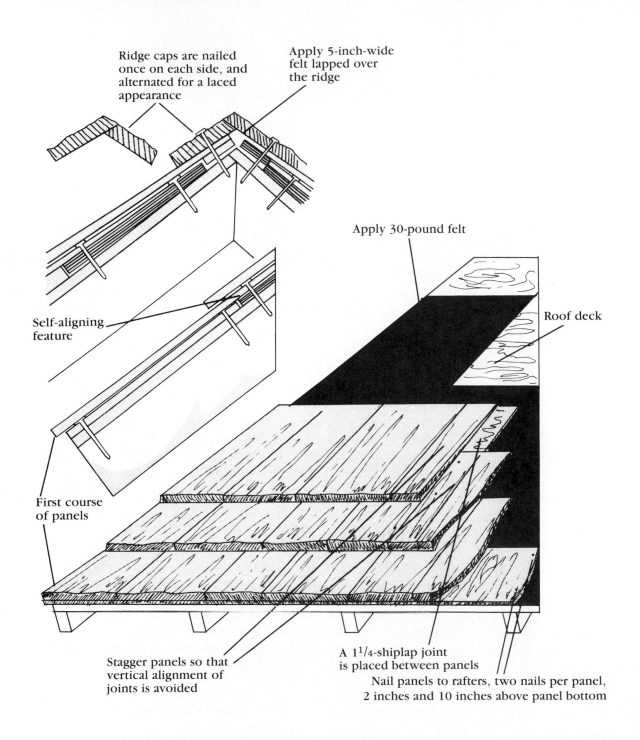

Ridge caps are nailed
once on each side, and
alternated for a laced
appearance

Apply 5-inch-wide
felt lapped over
the ridge

Apply 30-pound felt

Roof deck

Self-aligning
feature

First course
of panels

Stagger panels so that
vertical alignment of
joints is avoided

A 1¹/₄-shiplap joint
is placed between panels

Nail panels to rafters, two nails per panel,
2 inches and 10 inches above panel bottom

Fig. 9-16   Panelized roofing.

wood fibers. Woodruf comes in 12-×-48-inch panels that are 50 percent more dense than natural wood and resemble wood shakes when installed. It takes 32 to 36 Woodruf panels to cover 100 square feet. In comparison, it takes about 200 wood shakes or 80 asphalt shingles to cover the same 100 square feet of roof surface. If left uncoated, Woodruf panels weather to a natural gray cedar color.

Chapter **10**

Metal Roofing

Galvanized, ribbed sheet-metal panels, aluminum shakes and shingles, and coil-steel panels are among the most durable roofing materials available. Although such materials are more expensive to purchase initially than asphalt fiberglass-based shingles, metal roofing lasts longer than shingles. With proper maintenance, metal roofing will not have to be replaced for 50 to 75 years. During that time, asphalt fiberglass-based shingles will need to be replaced two to three times, depending on their weight.

While some sheet-metal materials require periodic painting, prepainted galvanized steel roofing is available in a wide variety of patterns and colors.

Aluminum Shakes and Shingles

Aluminum shakes and shingles vary in style and installation requirements. Detailed application instructions are provided by the manufacturers. The advantages of using aluminum roofing include long life, low maintenance, light weight, and attractive appearance. Because aluminum reflects up to 80 percent of the radiant heat from sunlight, your summer house-cooling requirements will be reduced. In winter, snow and ice are quickly shed from the roof as the warmed aluminum encourages any snow or ice to slide off the surface.

Aluminum shakes manufactured by Reinke Shakes, Inc. must be installed similar to wood shakes (see chapter 9). An 18-inch-wide interlay strip of 15-pound felt must be applied beneath each course of shakes. As with wood shakes, applying these interlay strips adds to the installation time.

Reinke shakes can be applied over new plywood, particleboard, chipboard, or similar sheathing material. The shakes can be applied over one layer of asphalt shingles or over one layer of wood shingles (not, however, wood shakes), but you'll first have to install suitable sheathing over the worn roofing material so that the shakes have a solid nailing base. See chapter 3 for guidelines on estimating the number of square feet of roof coverage.

To begin installing the shakes, nail down $2^1/_2$-×-$^1/_8$-inch furring strips from rake to rake and flush with the eaves. Because the factory-drilled nail holes are located on the exposed portion of the shakes, you'll have to install three layers of 15-pound felt with the shakes.

At the eaves, install two layers of 18-inch-wide felt. Overlap the second layer of felt 2 inches higher than the first layer (see Fig. 10-1). The interlay felt strip beneath each course of shakes makes up the third felt layer. The three felt layers seal around the nails, and moisture that penetrates the nail holes will flow over the top of the next course of shakes.

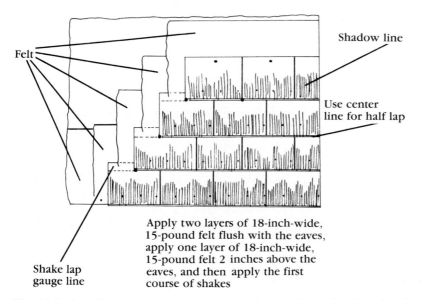

Fig. 10-1 Install two layers of 18-inch-wide felt over the roof deck and an interlay strip of felt with each course of shakes.

Working left to right, apply the starter row of shakes. To match the corrugations on the starter course, and therefore, subsequent courses of shakes, cut off the shadow line of one shake, turn the shake upside down, and lap half of the shake along the first shake being installed.

Install the interlay felt strip. Stagger the alignment of the first shake of the second course by half the width of a shake so that every other course of shakes is aligned.

There are two options for installing Reinke shakes along the rakes of a roof. One method is to install drip edge at the rakes (refer to chapter 3) and to apply the shakes either flush with the drip edge or overhanging the edge by 1 inch. The second method is to not use drip edge, but to position and then trim each shake so that it extends 1 inch over the rake. Bend down the rake edge of the shake about 1 inch and then nail the shake in place (Fig. 10-2). The edge is bent to make the rake watertight. Drip edge does the job better; it provides a straighter, more attractive appearance, is inexpensive, and requires less work to install.

To install shakes at a valley (Fig. 10-3), apply two layers of 15-pound, 36-inch-wide felt down the length of the valley. Install a length of aluminum valley material by driving nails only at the edges of the valley. See chapter 6. Apply silicone caulking under each shake as you install the shake courses at the valley. Using tin snips, trim the shakes along the corrugations. Use a power saw to cut across the shake corrugations.

To apply ridge or hip capping, install a folded 18-inch strip of felt (9 inches total width) along the hip or ridge. Use a utility knife to cut off the

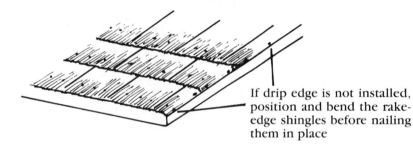

If drip edge is not installed, position and bend the rake-edge shingles before nailing them in place

Fig. 10-2 Position, bend, and install the shakes at the rakes. An alternative method is to use drip edge at the rakes.

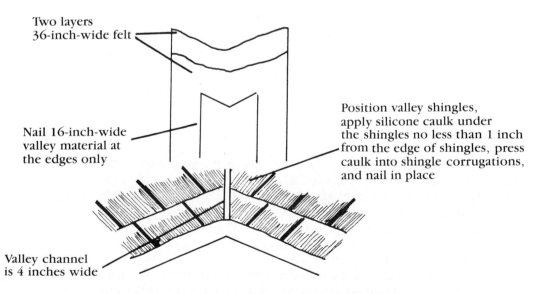

Two layers 36-inch-wide felt

Nail 16-inch-wide valley material at the edges only

Position valley shingles, apply silicone caulk under the shingles no less than 1 inch from the edge of shingles, press caulk into shingle corrugations, and nail in place

Valley channel is 4 inches wide

Fig. 10-3 Valley installation details.

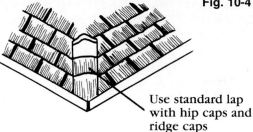

Fig. 10-4 Hip and ridge cap installation details.

Use standard lap
with hip caps and
ridge caps

shake shadow line to make the caps. Use the standard lap and coverage for the capping (Fig. 10-4).

Galvanized Sheet Metal

Most *galvanized sheet-metal roofing*, such as Channeldrain by Wheeling-Pittsburgh Steel, Wheeling Corrugating Co., has attractive ribbed seams. *Terne*-coated stainless steel, by Follansbee Steel Corporation, is a similar product. The steel sheets are coated with molten zinc in a process known as *hot-dip galvanizing*. The zinc provides corrosion resistance. For residential structures, the metal thickness should be about 26 gauge.

The first step before installation is to estimate the number of square feet of roof coverage (see chapter 3), the number of self-tapping sheet-metal screws with neoprene washers, and the number of 10-foot-long ridge caps and rake caps you will need.

The panels are installed vertically in sheets that stretch from the eaves halfway to the ridge of a typical residential home. Begin by placing the first panel at the corner of the roof away from the prevailing winds. Work from the eaves to the ridge. See Figs. 10-5 through 10-9 for the proper installation sequence.

Channeldrain's 3-foot-wide, 13-foot-long panels can be installed quickly once the first panel has been carefully and properly aligned and positioned at the eaves. Make sure that the first panel is square; the proper alignment of the remaining sheets depends upon the correct placement of the first panel.

Once the first panel is aligned, use a drill to install screws every 2 to 4 feet. Channeldrain panels are secured at the ridge with caps and at the rakes with end caps. At the eaves, the sheets are pulled together at the ribs and secured with short screws. Always set fasteners flush with the panel surface. Never overdrive nails or screws and do not dimple the steel. Subsequent sheets are installed quickly until you reach an obstacle such as a vent pipe, a chimney, or a valley.

To cut or trim panels, it is essential that you use an electric saw with a steel cutting blade or a Carborundum blade. Place the panel exterior side down so that the exterior surface will not be marred. Carefully brush off

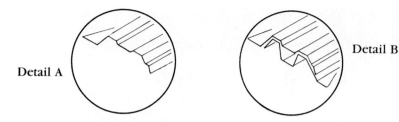

Detail A

Detail B

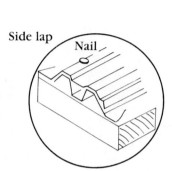

Side lap

Nail

Fig. 10-5 Detail A shows how panel 1 is aligned with the rake and eaves. Detail B shows how panel 1 and panel 3 are lapped. Side-lapped panels must be nailed so that the panels are securely drawn together. Do not overdrive the fasteners or dimple the steel.

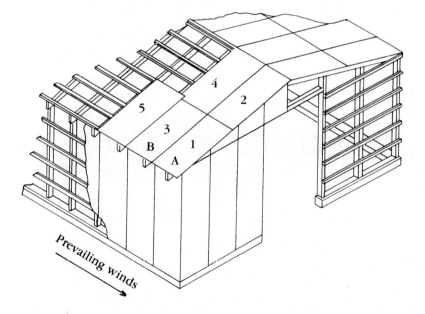

Prevailing winds

any metal particles and filings that otherwise could cause rust marks or bleeding on the installed panel surfaces.

For a vent pipe, cut a circular hole about $1/_2$ inch larger than the pipe, and make a horizontal slot long enough for the flange to fit. The "top" of the flange must slide under the sheet-metal panel. The "bottom" of the flange fits over the pipe and is installed on top of the panel (Fig. 10-10).

Fig. 10-6 Install the first panel with a 1-inch overlap at the eaves and the rake.

With middle-of-the-roof chimneys, you'll have to position and then cut at least two sheet-metal panels to fit around first one side and then the other side of the chimney. Rake chimneys are much easier to handle. Basically, all you will have to do is measure and cut the area to be removed

Fig. 10-7 Proper alignment of the first panel ensures that additional panels are correctly installed.

Fig. 10-8 Position the remaining panels to allow for the correct amount of overlap.

Fig. 10-9   Use tin snips to cut counterflashing. Flashing must be installed flush with the wall and on top of the sheet-metal panels.

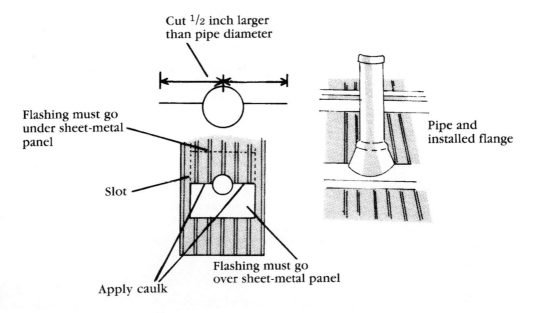

Fig. 10-10   Flange installation details.

from a panel, install the sheet, and flash the chimney area. See the section on chimneys in chapter 7.

Installing panels at valleys requires careful cutting and trimming to obtain a watertight, as well as a neat, appearance. Install the valley material as described in chapter 6, but note that no border shingles are required with sheet-metal roofing. The sheet-metal panels must extend over the valley surface. Do not drive screws closer than 3 inches from the edge of the valley material.

Stone-Coated Galvanized Steel Panels

At about 140 pounds per 100 square feet, Gerard steel roofing tiles or shake panels combine the durability of lightweight galvanized coil steel with an attractive, crushed-stone appearance. The panels must be attached to either steel or wood battens to create a class A fire-retardant, 50-year roof covering that can withstand winds of up to 120 miles per hour. The shakes or tiles can be applied to new construction, mansards, or over a layer of worn roofing materials (except clay or cement).

Gerard products are available for installation only through authorized roofing contractors. Refer to the Resources at the end of this book for the Gerard customer service address and telephone number.

Corrugated
Asphalt Roofing

Because asphalt impregnated, or *corrugated asphalt roofing*, sheets and tiles are lightweight, strong, and easy to install—the corrugations help make them almost self-aligning—Onduline roofing products (Fig. 11-1) can be a good choice for do-it-yourselfers. Available in various colors and in either a granulated or a smooth finish, the sheets and tiles have a class C fire rating, and can be painted by brush or by spraying with 100 percent acrylic latex exterior paint.

Onduline roofing sheets or tiles can be applied over a layer of felt on a new plywood roof deck or over a layer of worn asphalt shingles. If you are reroofing over metal roofing, Onduline recommends using sheet roofing rather than tiles. When applied over a layer of worn asphalt shingles, Onduline roofing needs no special preparations. If you have one layer of metal roofing with standing seams, you must flatten the seams before you install the sheets. With corrugated metal roofing, you must first apply wood cross strips, called *purlins,* no wider than 24 inches on center, as a nailing base (see Fig. 11-2).

Onduline sheets and tiles can be installed on a 3-in-12-foot-pitch or steeper slope. Sheets must be installed over purlin support strips that are typically spaced 24 inches on center. Tiles can be applied to a similar slope over a solid deck of plywood or over one layer of worn roofing. Special installation requirements, such as placing the purlins 18 inches on center, should be followed if your area receives unusually heavy snow accumulation. If your roof deck is more than 80 feet from the eaves to the ridge line, don't use sheets.

While the manufacturer suggests that Onduline roofing can be applied over two worn layers of asphalt shingles, keep in mind that most residential structures are not designed to support three layers of roofing materials.

Fig. 11-1 Onduline roofing products can be installed over one layer of worn roofing or on a new deck.

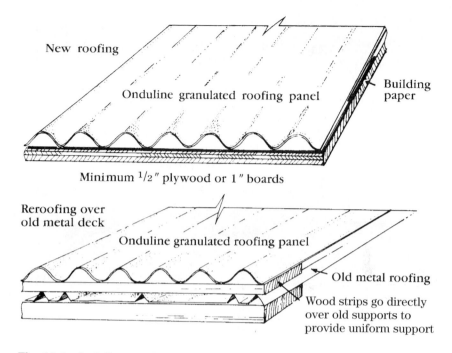

New roofing

Onduline granulated roofing panel

Building paper

Minimum $1/2''$ plywood or $1''$ boards

Reroofing over old metal deck

Onduline granulated roofing panel

Old metal roofing

Wood strips go directly over old supports to provide uniform support

Fig. 11-2 Onduline panels can be applied over felt on a new deck, where worn roofing has been removed, or over a layer of worn roofing. To reroof over a layer of metal roofing, first install purlins as a nailing base.

In addition, the nails used to install the new roofing must adequately penetrate the roof deck through more than two layers of shingles. According to Onduline, nails should penetrate 1 inch through the deck (see Fig. 11-3).

Installing Onduline roofing creates a dead air space between the new and old roofing material. This dead air space provides a barrier to the transfer of heat during the summer and additional insulation during the winter.

Tools you will need are a claw hammer or a roofing hatchet, a chalk line, a nail apron, a tape measure, a utility knife and blades, and an electric saw with a carbide-tipped blade. Use the utility knife to cut sheets or tiles parallel with the corrugations. Use the saw to cut across corrugations. Remember to wear safety eyeglasses when cutting across corrugations. A powered nail gun can be used to install tiles, but don't attempt to use a staple gun on either Onduline sheets or tiles.

Step-by-step illustrations and directions for applying sheets and tiles on rectangular sections, hips, valleys, out-of-square sections, and flashing (Fig. 11-4) and skylights (Fig. 11-5) are detailed in the installation brochures that come with the materials. Check with your local building suppliers and Onduline for more information on the availability of corrugated asphalt roofing products; the address is in the Resources list in the back of this book.

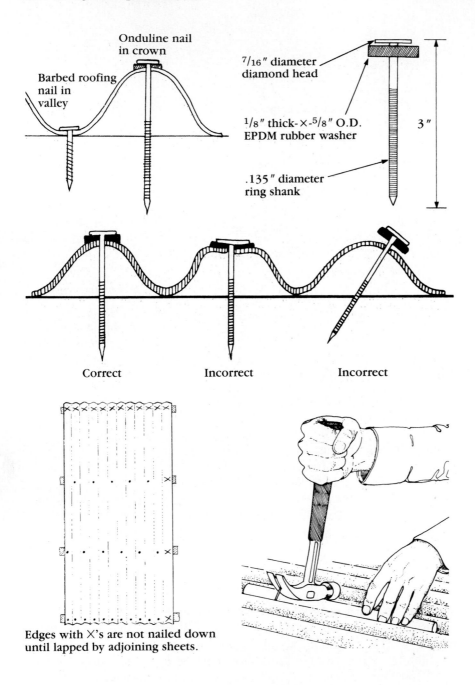

Edges with X's are not nailed down
until lapped by adjoining sheets.

Fig. 11-3 Drive barbed roofing nails through tile valleys 4¹/₂ inches from the top edge of the tile so that the courses lap the nail heads. Drive 3-inch Onduline nails through corrugation crowns at the eaves, at the ridge cap crowns, and at the lower corners and lower center of the tiles. If you use 4-inch nails, install up to 1¹/₂ inches of rigid insulation beneath Onduline roofing.

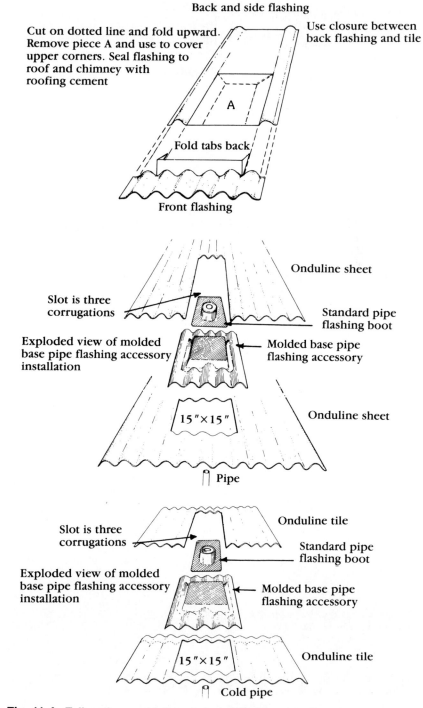

Back and side flashing

Cut on dotted line and fold upward. Remove piece A and use to cover upper corners. Seal flashing to roof and chimney with roofing cement

Use closure between back flashing and tile

A

Fold tabs back

Front flashing

Onduline sheet

Slot is three corrugations

Standard pipe flashing boot

Exploded view of molded base pipe flashing accessory installation

Molded base pipe flashing accessory

15″×15″

Onduline sheet

Pipe

Slot is three corrugations

Onduline tile

Standard pipe flashing boot

Exploded view of molded base pipe flashing accessory installation

Molded base pipe flashing accessory

15″×15″

Onduline tile

Cold pipe

Fig. 11-4 Follow these guidelines to install flashing at walls and around pipes. Refer to chapter 7 for more information on flashing roof obstacles.

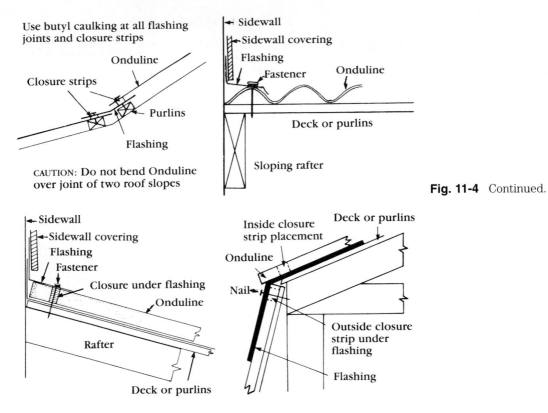

Use butyl caulking at all flashing joints and closure strips

Onduline

Closure strips

Purlins

Flashing

CAUTION: Do not bend Onduline over joint of two roof slopes

Sidewall

Sidewall covering

Flashing

Fastener

Onduline

Deck or purlins

Sloping rafter

Fig. 11-4   Continued.

Sidewall

Sidewall covering

Flashing

Fastener

Closure under flashing

Onduline

Rafter

Deck or purlins

Inside closure strip placement

Deck or purlins

Onduline

Nail

Outside closure strip under flashing

Flashing

Estimating Materials

Use the information outlined in chapter 3 and the Onduline guidebook to help you determine your roof's total number of square feet. Remember to

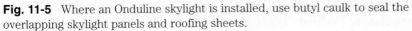

Fig. 11-5   Where an Onduline skylight is installed, use butyl caulk to seal the overlapping skylight panels and roofing sheets.

include 10 percent for waste. Consult with your local Onduline dealer for more details.

Sheets

Each sheet of Onduline roofing covers 26 square feet. Because the sheets must be overlapped when installed, it takes $4^1/_2$ sheets to cover 100 square feet of roof surface. You'll need 24 nails to install each sheet and 38 nails for each ridge cap.

To find the total number of closure strips you'll need, first determine the total length—in inches—of all the eaves and twice the length of the hips, valleys, and rakes. Divide this total by 44 (the length in inches of one closure strip) to obtain the number of strips needed.

Tiles

Each tile of Onduline roofing covers $6^1/_2$ square feet. Because tiles must be overlapped when installed, 24 tiles will cover 100 square feet of roof surface. To install each tile, you'll need 12 barbed roofing nails and 12 galvanized, ring-shanked, washered nails for each tile. Each ridge cap provides 6 feet of coverage. The total number of closure strips needed can be determined by finding the length, in inches, of all the eaves and twice the length of hips, valleys, and rakes. Divide the total by 44 (the length of one strip) to find the number of strips needed.

Applying Sheets

To install Onduline sheets, follow the guidelines described in chapters 3 and 4. Before nailing on the first sheet, install drip edge along all of the rakes and eaves. The next step is to establish a uniform overhang at the rakes and at the eaves. The sheets should be installed flush with the drip edge at the rakes, and the sheets should overhang the eaves by $1^3/_4$ inches.

If you are right-handed, start at the lower left corner of the back of your roof in order to keep the work in front of you. If you are left-handed, begin work at the lower right corner of the back of the roof. The following information is oriented for right-handed people. Reverse the starting points—to the opposite side of the roof sections—if you are left-handed.

Starter Course

At the left-hand corner of the rake and eaves, measure in 48 inches—remember to align the sheets with the drip edge—and make a mark. At the ridge, measure in 48 inches and make another mark. Snap a chalk line between these marks for proper sheet alignment. If you are working alone, you can loop the hook at the end of the chalk line over a nail head and then snap the chalk line. See Fig. 4-47.

To provide a $1^3/_4$-inch overhang at the eaves and at the left-hand corner of the rake and eaves, measure $77^1/_4$ inches up from the eaves. Measure

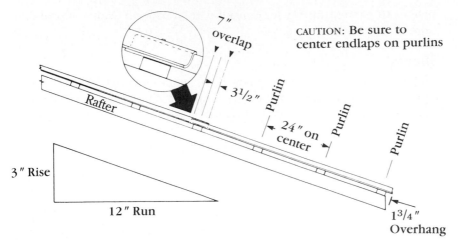

Fig. 11-6   The sheet courses must overlap by at least 7 inches.

up the same distance at the right-hand corner and snap a chalk line between the marks.

Position the first sheet of roofing at the left-hand corner of the rake and eaves. At the purlins along the rake, drive washered nails only through the top of corrugations. Do not install nails along the top or right-hand side of the first sheet until the next sheet is positioned, properly overlapped, and ready to be fastened. If you are using closure strips, install the first strip before you fasten the corrugations along the eaves. Now check for the correct alignment of the first sheet so that the remaining sheets are accurate.

Next, measure off a series of 44-inch marks at both the eaves and ridge line. Snapping chalk lines between these marks will provide for a 1-inch sidelap of corrugations as you install the remaining sheets for the first course at the eaves.

Subsequent Courses

The second and subsequent courses are installed so that they align with the lower courses or so that they overlap in the middle of the course below. To create an overlapping pattern, measure in halfway on a sheet, and cut down the length of the sheet between corrugations with a utility knife. Install the half sheet, as the first sheet of the second course at the left-hand rake, and save the other half sheet for the opposite rake. If you use the overlap pattern, measure and snap additional vertical chalk lines every 44 inches across the roof.

The second course must be lapped at least 7 inches over the top end of the first course (Fig. 11-6). So that the nailing pattern is secure, be certain that the center of each overlap is over the center of the purlin. Continue installing courses on both sides of the roof until you reach the ridge line (Fig. 11-7).

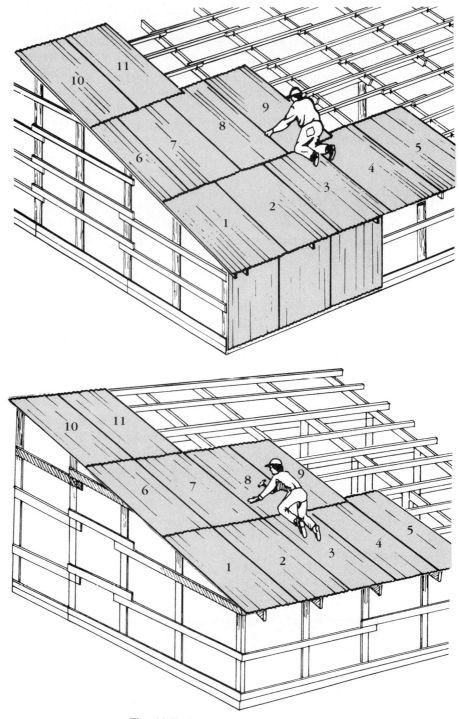

Fig. 11-7 The sheet installation pattern.

At the ridge, the top ends of the sheets from both sides of the roof should be within 2 inches of each other in order to support the ridge capping. Instead of installing ridge capping, consider adding a continuous ridge ventilator (chapter 13). If you do use ridge capping, position the first cap away from prevailing winds and 3 to 6 inches from the ridge end. Install a closure strip and drive washered nails through the cap, the closure, and the underlying sheets. Next, cut the ridge cap portion that projects from the rake so that the resulting fold provides a weather guard. Install the remaining ridge caps so that they overlap by 7 inches.

Applying Tiles

Before you install Onduline tiles, read the roof preparation guidelines described in chapters 3 and 4. Before installing the first tile, nail drip edge along all the eaves and rakes. Allow for a 1-inch overhang at the eaves and plan to align the tiles flush with the drip edge at the rakes.

If you are right-handed, begin at the lower, left corner of the back of the roof in order to keep the work in front of you. If you are left-handed, start working at the lower right corner of the back of the roof. In order to simplify the instructions, the following descriptions are oriented for right-handed persons. If you are left-handed, reverse the starting points to the opposite sides of the roof sections.

Starter Course

At the left-hand corner of the rake and eaves, measure in 48 inches and make a mark. Remember to allow for the tiles to be aligned with the drip edge. At the ridge, measure in 48 inches and make another mark. To ensure proper tile alignment, snap a chalk line between the marks. If you are

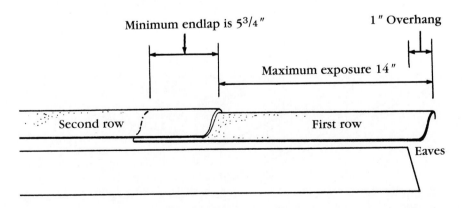

Fig. 11-8 Tile installation at the rake and eaves.

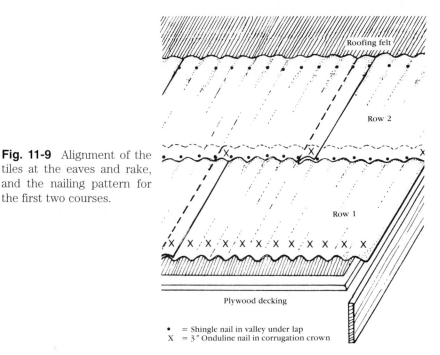

Fig. 11-9 Alignment of the tiles at the eaves and rake, and the nailing pattern for the first two courses.

installing the tiles by yourself, hook the end of the chalk line over a nail head and then snap the chalk line. See Fig. 4-47.

To allow for a 1-inch overhang at the eaves, measure 78 inches up from the left-hand corner of the eaves and rake. Measure up the same distance at the opposite end of the roof section and snap a chalk line between the marks.

Position the first roofing tile at the left-hand corner of the rake and eaves. At the purlins along the rake, drive washered nails only through the crown of the first corrugation so that the edge is aligned with the rake. Position a closure strip between the drip edge and the tile (Fig. 11-8). Using barbed roofing nails positioned $4^1/_2$ inches from the top of the tile, nail the tile between the corrugation. Now is the time to make sure the first tile has been installed properly. Correct alignment of subsequent tiles depends on proper installation of the first unit. See Fig. 11-9.

Measure off a series of 44-inch marks at both the ridge line and at the eaves. Snap chalk lines between these marks to ensure a 1-inch sidelap (Fig. 11-10) of corrugations for the subsequent tiles. Allowing for a 1-inch sidelap, install all but five of the remaining tiles of the first course. Position, but do not nail down, the remaining five tiles at the eaves in order to determine whether or not the last tile in the course will align with a "valley" between tile corrugations. Adjust the amount of sidelap for the last five tiles by changing the sidelaps until the proper alignment at the rake is achieved (Fig. 11-11). When the alignment is correct, nail the last five tiles in place.

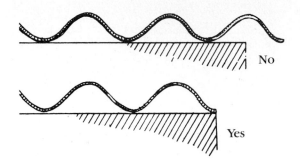

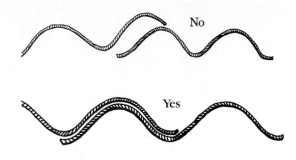

Fig. 11-10 The corrugations must overlap by at least 1 inch.

Fig. 11-11 Adjust the sidelap for proper alignment at the rake.

Subsequent Courses

The second and subsequent courses must be installed so that they overlap the middle of the course below. To create an overlapping pattern, measure in 14 inches from the rake at the left-hand edge of the first course. Count off six corrugated crowns on the first tile of the second course (Fig. 11-12). Cut down the length of the tile between corrugations with a utility knife,

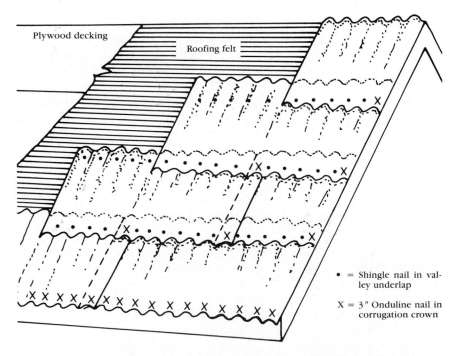

Plywood decking

Roofing felt

• = Shingle nail in valley underlap

X = 3″ Onduline nail in corrugation crown

Fig. 11-12 Note the nailing patterns and the overlapping of the courses.

and install the half tile as the first tile of the second course at the left-hand rake. Save the other half of the tile so that it can be installed at the opposite rake. Using the first half tile as a starting point, measure and snap additional vertical chalk lines every 44 inches across the roof section.

The second course must be lapped at least 14 inches over the top end of the first course. Continue installing courses on both sides of the roof until you reach the ridge line.

At the ridge, the top ends of the sheets, from both sides of the roof, must be within 2 inches of each other. Consider adding a continuous ridge ventilator (chapter 13) instead of ridge caps (Fig. 11-13). If you do use ridge capping, position the first cap away from prevailing winds and 3 to 6 inches from the ridge end. Install a closure strip and drive washered nails through the cap, the closure, and the underlying tiles.

To provide a weather guard at the rake, cut and fold the ridge cap portion that projects from the rake (Fig. 11-14). Install the remaining ridge caps so that they overlap by 7 inches.

Hips and Valleys

For hip roofs, courses of Onduline sheets or tiles must be trimmed along each side of the hip so that the courses support the ridge capping installed over the hip. To determine if you have the proper coverage at the hip, measure the width of the capping, divide by two, and measure the result from each side of the center of the hip. Use these marks to snap chalk lines on each side of the hip.

Before installing the hip capping, make certain that you have adequate framing support as a nailing base along the hip and at the ridge line. Cut two-corrugation-wide closure strips, and use butyl caulking to install a two-corrugation-wide overlapping pattern (Fig. 11-15). Install a 7-inch endlap at the bottom of the hip.

For valley flashing, nail in place a minimum of 36-inch-wide sheets. If your valley flashing is not long enough to cover the full length of the valley, allow 6 inches for each overlap between sections. Refer to chapters 6 and 7 for additional information on valleys and other roof obstacles.

Position and trim, but to not nail in place, the Onduline sheets or tiles at an angle to allow for a 6-inch-wide center channel. Turn over the cut sheets or tiles and use butyl caulk to attach two-corrugation-wide pieces of closure strip in an overlapping pattern. See Fig. 11-16. Add a $1/2$-inch bead of caulk on the flat side of the closure strips and on the sheet between the closure strips. Turning the sheets or tiles upright, position the roofing so that the caulked closures meet the valley flashing. Now nail the sheets or tiles in place (Fig. 11-16).

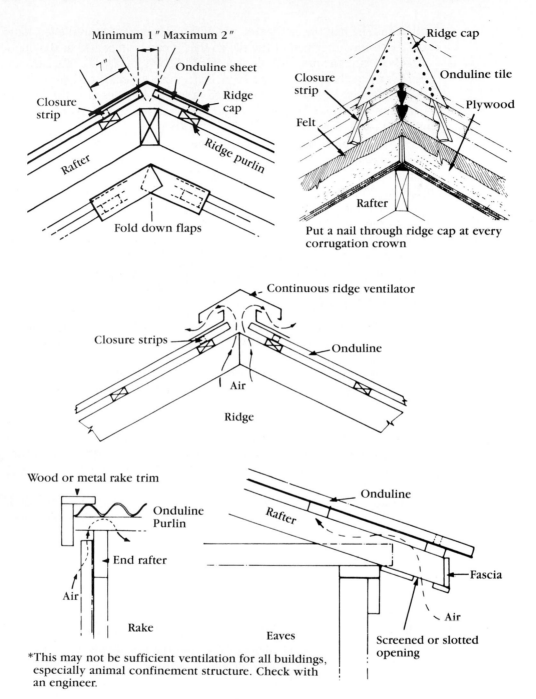

Fig. 11-13   Ridge line ventilation details.

Fig. 11-14 Closure strips and ridge cap installation provide protection from wind-driven moisture at the roof edges.

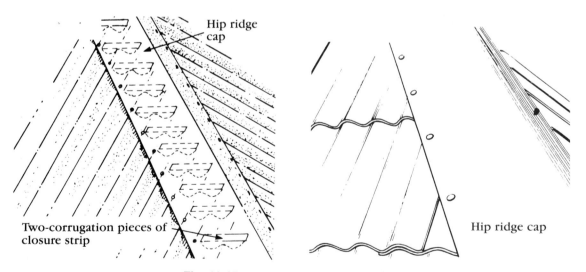

Fig. 11-15 Hip capping installation details.

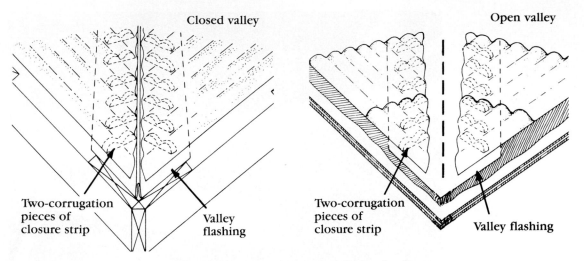

Closed valley

Open valley

Two-corrugation
pieces of
closure strip

Valley
flashing

Two-corrugation
pieces of
closure strip

Valley flashing

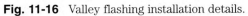

Fig. 11-16   Valley flashing installation details.

Tile Roofing

Of all the popular roofing options commonly available, tile roofing requires some of the most complicated choices. Application techniques for *concrete tile roofing*, *clay tile roofing*, and composite tile roofing vary with specific manufacturers' products. Therefore, this chapter provides general guidance based on a "typical" reroofing project: that is worn materials were torn from the roof, damaged sheathing was replaced, a double layer of felt was installed, and concrete tile was applied. No matter what type of materials you select, remember that the essential component for a long-lasting, waterproof tile roof is the underlayment.

Traditional and new concrete or hybrid clay tile products vary considerably in appearance, style, weight, cost, life span, and installation techniques. Refer to chapter 1 for a comparison of various roofing materials.

Roof tile requires expensive materials, renting or buying expensive equipment, hiring expensive labor, or possessing above-average roofing knowledge and skills. You'll also have to contend with moving heavyweight roofing materials. While individual tiles usually are not at all difficult to lift, a typical residential home might require more than 16 tons of material to cover the roof. In addition, you'll be installing two separate roof systems on your home. Layers of underlayments and flashings that waterproof the roof surface must be added and then the tiles themselves.

While tile certainly provides a distinctive appearance and will divert some of the climatic elements a roof is subjected to, tiles alone are not designed or intended to waterproof a roof. Tile roofing advertised to last a "lifetime" should easily weather 50 years or more, but an inadequate underlayment of 15-pound felt could wear out in a decade. Be especially wary of tract home developers who skimp on roof underlayments for hundreds or even thousands of their new homes while promoting the advantages of tile.

When that single layer of 15- or 30-pound felt wears out, someone will have to lift your "lifetime" tile, strip the battens and the worn underlayment, replace and upgrade the waterproof barrier, replace the battens, and reinstall the tile.

Don't neglect the importance of applying several layers of felt. A second layer of felt or roll roofing will not add substantially to the overall cost of installing a tile roof. For example, consider installing two layers of 30-pound felt or a layer of 45-pound felt and heavy roll roofing. Also, be sure to use quality flashings and fasteners so that your roof system will truly last a lifetime.

Because of tile roofing's substantial weight per 100 square feet, homeowners must have a building engineer inspect the structure to determine if the house needs reinforcement or sheathing replacement. Braced purlins at mid-span could add hundreds of dollars to the cost of your roofing project. A lightweight concrete shake or tile or one of the newest hybrid fiber-cement roofing products are alternatives to bracing the house frame to accommodate heavyweight roofing products. However, some manufacturers of the newer roofing materials require that their products be installed only by authorized roofing applicators. You'll have to calculate and compare the potentially higher costs of various heavyweight and lightweight materials versus the potential costs of reinforcing the house frame.

Cement, concrete, and composite products are made to resemble tiles or slates. Cement shakes are manufactured to imitate wood shakes. When comparing product brochures and actual samples, keep in mind that efflorescence and hue variance can occur. To avoid problems regarding color selection, inspect samples when you order your product, then confirm your product style and color selection when the materials are delivered.

Composite products combine Portland cement and wood fiber to provide class A fire ratings, limited 50-year warranties, and relatively lightweight materials. Concrete tiles weigh about 900 to 1200 pounds per 100 square feet and provide a class A fire rating.

Because composite materials are less brittle than concrete or clay, composites are less likely to break, chip, or crack during transport and installation. Because they break easily, avoid walking on installed concrete tiles. Composite roofing is usually walkable once installed on a suitably sloped roof.

Natural clay products can vary in color. To avoid potential undesirable color patterns on the finished roof, periodically inspect the clay roof tiles from the ground at a distance of about 50 feet. Periodic inspections from the ground can reveal the best way to obtain a random color sequence, ensure that colors are blended appropriately, and avoid streaks, stair-stepping, or checker-boarding.

A raised fascia board, special under-eaves tiles, or birdstops (for mission tiles) must be installed at the eaves before the first course of clay tile is installed. Cement tile roofs require a raised fascia and an anti-ponding metal flashing strip at the eaves so that the first course of tiles is installed at the same pitch as the *field* courses. Along with the watertight underlayment, horizontal lathing—installed 14 inches apart—must be nailed to the roof deck. Some tile installation techniques and products require vertical battens as well as horizontal lathing.

Material Delivery

Once you have placed your order for the tile you have selected, you must carefully plan and coordinate tearing off the worn roofing materials, making repairs to sheathing, and drying in the roof surface. If a contractor will be tearing off the old roof, discuss who will install the new layers of felt and how soon the felt will be applied after tear-off work is completed. You don't want your roof exposed to the elements for a week while you wait for materials to be delivered.

Asking the driver for a brief inspection of the materials while they are still on the delivery truck could save you considerable delays and time and effort. Open a bundle of tiles so that you can determine if the tile exactly matches the brand, style, weight, and color (beware of mixed blend numbers) you ordered from the supplier.

Because of the considerable volume and weight of the tile delivered to the job site, materials frequently are not immediately lifted to the roof of the building. Unless you have made prior arrangements with the supplier, the delivery driver generally will not want to wait the several hours it will take you and several helpers to distribute short stacks of individual tiles evenly across the roof surface. Instead, the supplier will unload pallets from the trailer using a forklift brought along specifically for that purpose (Fig. 12-1). Be certain to ask your supplier if rooftop delivery is an option. Some companies will provide this service for a flat rate or an additional fee of several dollars per every 100 square feet of material purchased.

The ideal storage method for factory-wrapped bundles is to stack them on wood pallets in a driveway or—space permitting—at the roadside. If you will not be lifting the bundles onto the roof immediately with the aid of a forklift you have rented (Fig. 12-2), cover the tiles with a tarpaulin. Keeping the paper wrappings dry makes carrying the bundles less difficult than if they are wet. The wrappings on wet bundles will split when you try to lift the bundles.

While individual cement tiles are easy to handle, don't attempt to carry the hundreds of stacks of tiles needed to stock a roof unless you are free of back troubles, have excellent health, and are in excellent condition. Wear gloves when unloading the tiles.

Installing the Underlayment

Techniques for installing the underlayment of a tile roof vary with the choice of one or more layers of felt or a combination of felt and roll roofing.

Fig. 12-1 Due to its considerable weight per 100 square feet, tile usually is delivered to the job site by a tractor-trailer. In this case, a forklift for moving the pallets of bundled tile also is transported to the location.

Fig. 12-2 A driveway or roadside space near the project is needed for temporary storage of pallets. If your supplier does not make rooftop deliveries, rent a forklift to stock the roof.

Double-lapped courses of felt should overlap by about 17 inches. Refer to chapter 4 for basic techniques for laying felt and chapter 14 for examples of roof-deck sheathing repairs.

With repairs to the roof deck complete and anti-ponding flashing installed at the eaves, a starter course of felt or roll roofing is applied at the eaves by cutting, in this example, a roll of 30-pound felt lengthwise. Use a hammer tacker to staple an identical strip of felt over the first layer (Figs. 12-3 and 12-4).

At the corner rake and eaves (Fig. 12-5), staple the top corner of the roll and begin rolling out a full course of felt so that the felt overlaps the rake and the starter course. Every 10 or 12 feet, adjust the felt (Fig. 12-6) to remove any wrinkles and to align the course even with the eaves.

At the opposite rake or wall, use a utility knife or tin snips to trim the felt so that the felt overlaps the rake or is even with the wall (Fig. 12-7). Begin a new course (Fig. 12-8) of felt by overlapping the first course by about 17 inches. If a full run of felt does not reach the rake, overlap the sections (Fig. 12-9) by about 8 inches. Be careful not to align consecutive vertical overlaps.

Use a utility knife to cut felt around obstacles such as pipes and vents (Figs. 12-10 and 12-11). With a double layer of felt, you must cut holes around obstacles for each layer of felt (Figs. 12-12 and 12-13). Use roofing

Fig. 12-3 To install a starter course at the eaves, cut two 17-inch horizontal strips of felt long enough to reach about the midpoint of the roof section. Staple the double layer even with the eaves and the wall.

Fig. 12-4   Roll out the two layers of felt halfway across the roof, use a stapler or roof nails to attach the double layer, and then repeat the process from the opposite direction. Be certain to overlap the midpoint by 6 or 8 inches.

nails or round- or square-headed *felt nails* to secure the bottom edge of each course of felt (Fig. 12-14). Drive in nails about every 3 or 4 inches across the overlapping layers of felt. Continue overlapping runs (Fig. 12-15) and applying courses (Fig. 12-16) until you reach the ridge line. At the roof ridge lines, overlap the opposite roof-section course with at least 6 inches of felt.

Install felt in the contours of a valley using the same techniques for waterproofing square or rectangular sections of the roof. First, determine what portion of the valley will be exposed to the weather. Next, install overlapping courses of felt (Figs. 12-17 through 12-21) so that the valley will drain water into the gutters and the tile can be applied along a straight course. If necessary, install short courses of felt and, eventually, tile. Use a chalk line as necessary for proper alignment of felt courses.

Where a chimney intrudes on the roof surface or along a rake, the layers of felt must go under the chimney flashing (Figs. 12-22 through 12-28). Felt is installed at the rakes so that it overlaps the edge of the rake by several inches (Fig. 12-29). The felt usually can be stapled directly to the rake edge or fascia. Keep in mind that the overlapping felt won't be visible because the rake will eventually be covered by a course of rake tiles.

Fig. 12-5 Overlap the starter course by using a full roll of felt. To provide protection against wind-blown moisture, staple the felt overlap at the eaves to the fascia.

Fig. 12-6 About every 10 or 12 feet, pull the felt taut. Make certain the felt is even with the eaves and that there are no wrinkles.

Fig. 12-7 At the wall, insert the felt under the flashing or weave the felt and step flashing to waterproof the surface.

Fig. 12-8 Begin a second course of felt by overlapping the first course by about 17 inches.

Fig. 12-9 Where a full roll of felt does not reach the rake, overlap sections by about 6 or 8 inches.

Fig. 12-10 Use a utility knife to cut felt around obstacles such as vent pipes.

Fig. 12-11   Make cuts around intrusions as close to the obstacles as possible.

Fig. 12-12   Continue installing the felt toward the wall.

Fig. 12-13 Install the second of the two layers of felt over the vent pipes.

Fig. 12-14 Once felt is properly aligned, use roofing nails or felt nails to continue securing the material.

Fig. 12-15 Avoid aligning consecutive overlapping vertical courses and trimming felt so that it overlaps at the eaves.

Fig. 12-16 Applying a double layer of 45-pound felt extends the life span of the underlayment and provide a durable, waterproof base for the tile.

Fig. 12-17 Applying felt over a valley that has been coated with polyurethane foam.

Fig. 12-18 The first course of felt at a foam-coated "dead" valley is positioned so that the felt extends well into the channel.

Fig. 12-19 The second full course of felt is installed along the edge of the valley.

Fig. 12-20 Dead valleys are unusual architectural features even for a region with very low annual rainfall. On this Phoenix, Arizona home, the valleys were constructed as part of additions to the original building.

Fig. 12-21   Lapping felt layers provides about 17 inches of double coverage.

Preparation

Snapping Chalk Lines

To install straight, evenly spaced courses of tile, horizontal chalk lines are snapped about every 14 inches (Figs. 12-30 through 12-32). Exact measurements vary with different tile products and the dimensions of the roof. Consult the tile manufacturer's course-spacing table for the product you are installing.

To obtain the proper course width, measure the length of the roof from the eaves to the ridge line. As you calculate where to lay your chalk lines and battens, remember that you'll have to adjust the spacing between courses to obtain at least the standard or minimum weather exposure for your tile. Plan to space your chalk lines and attach your battens so that you avoid a short or odd-shaped course of tile at the ridge.

Installing Battens

Once the chalk lines have been snapped, you must lay out the battens along the chalk lines and then nail them in place, preferably with a pneumatic stapler or coil nailer (Fig 12-33). An alternative method to attach battens par-

Fig. 12-22 A short course of felt abuts the chimney and valley.

Fig. 12-23 The felt overlaps the ridge line and extends toward a vent pipe and chimney.

Fig. 12-24 At the chimney, cut the felt to fit around the obstacle.

Fig. 12-25 To ensure a waterproof chimney, fit the edge of the felt under a portion of the step flashing.

Fig. 12-26 Make certain the layer of felt fits securely under the step flashing.

Fig. 12-27 Apply felt under the flashing at the rear of the chimney.

Fig. 12-28 At the opposite side of the chimney, install the first course of felt so that the amount of valley exposure is equal on both sides of the chimney.

allel to the roof rafters is to nail horizontal strips of lath over the battens to create a raised grid on which to hang the tile. Raising the tile above the felt provides a dead air space and permits moisture to reach the felt to drain toward the gutters. When all the battens are installed, you are ready to stock the roof with tile.

Stocking the Roof

Because of its considerable 100-per-square-foot weight and the volume of material needed, carrying roof tile up a ladder is not practical. Also, because of the weight, you must avoid stacking too much tile at one location on the roof surface. Once you place the tile on the roof surface, you don't want to move individual tiles several times while you work.

If your materials supplier does not offer rooftop delivery, consider renting a forklift and—depending on how much help you have—plan on a day to stock your roof (Figs. 12-34 and 12-35). Beginning at the third row of battens up from the eaves, stock the roof with stacks of tiles seven or eight tiles high, distributed about 6 to 8 inches apart. Place rake tiles so that they are easily accessible near the rakes and distribute ridge-cap tiles near the ridge line and any hips. Additional tile intended to replace material that will inevitably be broken during installation should be stacked near the ridge line. A small percentage of replacement tile can remain on the ground.

Fig. 12-29 At the rakes, staple several inches of overhanging felt to the fascia. You can add felt nails to better secure the felt to the fascia for protection against wind-blown moisture. The felt will not be visible once a course of rake-edge tiles is installed.

Fig. 12-30 Roof sections must be measured and marked to determine the best spacing for the tile courses.

Fig. 12-31 To avoid a short course of tile near the ridge line, space the courses evenly up the roof surface. Keep in mind the manufacturer's minimum weather exposure for the product you are installing on your roof.

Fig. 12-32 Snap a series of parallel chalk lines about 14 inches apart. Exact dimensions depend on the type of tile you use and the length of the roof section.

Fig. 12-33 Battens on which the tile will be hung are nailed parallel to the chalk lines.

Fig. 12-34 Once all battens are installed, raise a pallet of bundled tiles to the rooftop.

Fig. 12-35 The roof is stocked with rows of tile about 6 inches apart and about seven or eight tiles high. The two rows nearest the eaves must remain open and "extra" tile is placed near the ridge line.

Laying Field Tiles

At a corner of the eaves and a rake or wall, begin the first course of tiles by installing a full tile flush with the flashing at the wall or about 1 inch in from the edge of the rake. The next tile interlocks with the first piece along a common *waterlock* edge (Fig. 12-36).

An individual tile hangs or hooks onto a batten by a lug that is part of the tile design and is fastened by hand with 8d galvanized nails through at least one of the manufacturer's three-per-tile, factory-drilled holes. Check your product installation instructions for guidance on the number of nails required to properly secure your tile. Product manufacturers, local building codes, and roofing contractors usually require that the first three courses up from the eaves, three courses in from any rakes, and the three courses down from the ridge line must be fastened with nails.

In most locations, the interior field tiles are held in place only by the lugs and gravity. If your home is located in a high-wind area or a seismic zone, nail all the tiles in place. Check your building code for compliance with local regulations.

Using a pneumatic coil nailer to fasten tile is not recommended because of the difficulty of positioning the nailer precisely above the drilled holes. In addition, the nails must not be driven too tightly or the tile might crack or break. If the nail head is not driven flush with the tile, the next course will not lie flat and the corner of the tile will break if you step on it.

The remaining tiles in the eaves course should be positioned loosely—with about a $1/16$-inch gap between pieces—to allow for the expansion and contraction of the material (Fig. 12-37). This gap can be adjusted as needed to permit an even number of tiles to fit the length of the first course so that the course ends at the rake or wall with a whole tile or a half tile.

Tiles should lie flat against a neighboring tile's rib or the batten. You can tell when a tile's waterlock isn't snug because it will rattle when tapped with your foot or hatchet handle. Make sure all waterlocks are snug and that individual tiles are properly aligned within the course.

With the first course positioned satisfactorily along the eaves, nail down the eaves tiles (Fig. 12-38).

Laying Subsequent Courses

Begin the second course of tile with a half tile so that consecutive courses do not align. The second course hangs on the second row of battens and is positioned so that the tiles cover at least the nail line of the first course (Fig. 12-39). The third course begins with a full tile so that it is offset from the second course (Fig. 12-40).

Remember to install rake tiles as you complete the courses at the eaves. Otherwise, you will have to risk breaking tile as you stand on the courses while adding the rake tiles. Secure the second and third courses of tiles with nails. For roofs that do not require all field tiles to be nailed, you can hang two courses of tiles at once (Fig. 12-41). Continue to offset consecutive courses and allow for the expansion gap between individual tiles.

As your installed courses approach the ridge line (Figs. 12-42 and 12-43), be certain to secure the last three courses with nails. If you have stocked your roof properly, you should have the correct amount of tile within reach to finish the roof section.

Roofing Around Obstacles

Valleys, angled roof sections, hips, walls, vent pipes, and other intrusions on the roof surface must be waterproofed with flashing and tile must be cut, broken, or trimmed to fit around the obstacles. For example, the bottom course of tiles that intersects a valley might have to be trimmed so that the tiles are positioned as a short course gradually tapers exposure into the valley (Figs. 12-44, 12-45, and 12-55).

Fig. 12-36 The first tile is installed flush with the wall flashing and even with the eaves. The second tile is aligned with the eaves and interlocks with the first tile. Both tiles are hung on the first batten and secured by 8d galvanized nails driven through factory-punched holes.

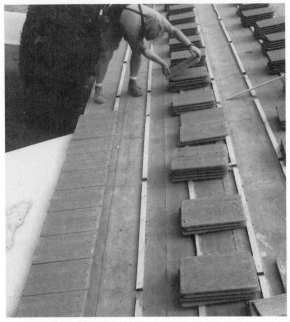

Fig. 12-37 The first-course tiles are hung on the battens but not nailed until the spacing between individual tiles is correct.

Fig. 12-38 After the $^{1}/_{16}$-to-$^{1}/_{32}$-inch gaps between first-course tiles result in a satisfactory appearance at the rake, nail the tiles in position.

Fig. 12-39 Position the second-course tiles.

Fig. 12-40 Position the third-course tiles.

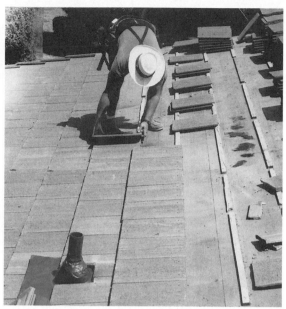

Fig. 12-41 Once past the first three courses, position two courses of tile at a time. On this roof, the field tiles from the fourth course in are not nailed.

Fig. 12-42 Within three courses of the ridge line, nail tiles to the battens.

Fig. 12-43 If the roof is properly stocked, as you approach the ridge line, you should have just enough material to tile the roof.

The base of a lead vent-pipe cover should be embedded in mastic (Fig. 12-45). Tiles must be cut to fit around a pipe (Figs 12-46 through 12-55). You can use a Carborundum blade on a portable circular saw, but cutting cement will quickly wear out the blade. Professional roofers use a diamond-tipped blade and a portable, gas-powered saw. Wear safety goggles and a dust mask when you cut cement tile.

For additional waterproofing, a second flange—spray painted to match the color of the tile—is installed over the lead jack. As with any roof vent flashing, at least one course of tile must go under the bottom of the flange while the next course of tile is positioned over the top edge of the vent flange (Fig. 12-51).

Where an angled roof section or a wall intersects tile courses, the intrusion must be counterflashed (Figs. 12-56 and 12-57). If necessary, partial or trimmed tiles can be inserted where you conclude a run at a wall or to continue the tile's standard pattern of overlapping courses past the intrusion (Figs. 12-58 through 12-61).

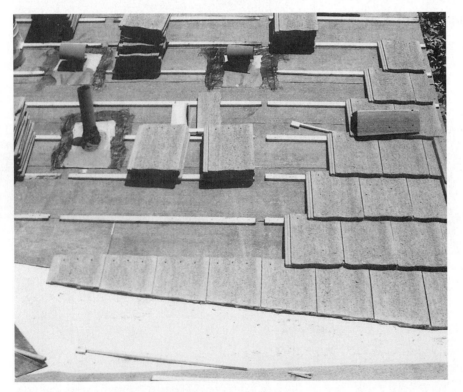

Fig. 12-44 Because the valley intersects the roof section at an angle, trim a short course of tile and install it so that the tiles gradually taper at the bottom and provide a straightedge for subsequent courses.

Fig. 12-45 Cover vent pipes with lead jacks embedded in mastic.

Fig. 12-46 Position tiles and mark to cut around the vent pipe.

Fig. 12-47   Making a vertical cut in the tile.

Fig. 12-48   Making a horizontal cut in the tile.

Fig. 12-49 Position the vent collar over the vent pipe. Position cut tile. Spray paint collars to match the color of the roof tiles.

Fig. 12-50 Position the courses below the vent pipe.

Fig. 12-51   Install the bottom of the vent collar over at least one course of tile.

Fig. 12-52   Install the top of the vent collar under at least one course of tile.

Fig. 12-53   Roofing cement holds the tile in position and helps to seal the surface around the vent.

Fig. 12-54   Seal the underside of the collar with roof cement.

Fig. 12-55   The weight from temporarily placing a tile on top of the collar helps secure the roof-cement seal. Nails are not used to fasten the front of the collar.

Installing Ridge Capping and Rake Tiles

Whenever possible, avoid walking on newly installed tile. Always try to attach ridge capping and rake tiles as you progress with the courses on each section of the roof. If you leave capping and rake tiles until last, you will undoubtedly break several tiles in the process of walking and working on the installed tile. Replacing broken tiles is an annoying task.

Along with securing ridge caps with nails, apply a spot of silicone adhesive to the overlapping portion of the underlying cap (Fig. 12-62). Ridge caps are overlapped just enough to cover the factory-drilled nail holes (Fig. 12-63). At the rakes, nail field tiles in position, about 1 inch from the edge (Fig. 12-64). Cover the somewhat ragged-looking rake edge (Fig. 12-65) neatly with rake tiles by nailing them to the fascia (Fig. 12-66). Some cement tile manufacturers produce special rake tiles that have a slightly different appearance than ridge capping tiles.

To fill in partial tiles at a rake, professional roofers (Fig. 12-67) gently strike the back of the cement tile with a hatchet so that it breaks evenly. With practice, breaking tile becomes routine, but the smaller the area to be trimmed, the more skill and luck it takes to obtain a clean break.

Fig. 12-56 Installing new siding to help counterflash the roof surface where the felt, flashing, and wall meet.

Fig. 12-57 Use flashing to seal the intrusion where a wall intersects the roofline.

Fig. 12-58   Fill in full and partial tiles against the wall flashing and around the intrusion.

Fig. 12-59   A full course of tiles extends past the roofline intrusion.

Fig. 12-60 Install a partial tile at the rake.

Fig. 12-61 Counterflashing, roofing cement, and tiles combine to waterproof the roofline intrusion.

Fig. 12-62   To help hold ridge capping in place, apply a dab of silicone adhesive before nailing each cap.

Fig. 12-63   Install ridge capping with the maximum amount of exposure to the weather.

Fig. 12-64 Installing a full tile and a half tile on alternating courses at the rake.

Fig. 12-65 Properly positioned rake tiles provide protection against wind-blown moisture.

Fig. 12-66 At the rake, position field tiles about 1 inch from the rake edge. The rake tiles are designed to cover portions of the tile and the fascia.

Fig. 12-67 With practice, a professional roofer can use a hatchet to break cement tiles to fit irregular spaces.

Fig. 12-68 Nail rake tiles directly to the fascia.

Fig. 12-69 Where the ridge and the rake meet, position and mark a rake tile to be trimmed.

Fig. 12-70 Trim the rake tile to form an end piece.

Fig. 12-71 Use silicone adhesive and nails to fasten the end piece.

Install rake tiles up both rakes where the rakes and ridge line converge (Fig. 12-68, page 303), then install the final ridge caps and end piece. Cut the end piece by first positioning and marking a rake tile (Fig. 12-69, page 303) and then trimming both ends to fit (Fig. 12-70). With the end piece properly positioned, nail the tile to the fascia (Fig. 12-71).

Chapter **13**

Ventilation

Proper attic ventilation can lower the cost of cooling your home by 60 percent, prolong the life of roofing materials, and prevent the premature deterioration of attic-insulation materials. Some 70 percent of American homes are not designed and constructed with proper ventilation in mind. Fortunately, installing wind-driven turbines, powered ventilators, roof-line louvers, cupolas, ridge vents, skylights, roof windows, and whole-house fans are easy do-it-yourself projects. The ideal time to install ventilating units is during the shingling of a new roof.

A Balanced System

When hot air rises to your home's attic, it will remain trapped there unless you have a good attic-ventilation system (Fig. 13-1). Without adequate ventilation, summer weather can easily superheat attic air to 140 degrees. Through convection, the heated attic air will raise the temperature of your living areas. Trapped, overheated air will radiate warmth downward along the ceiling, walls, and joists. The super-heated attic air keeps warm air in your living areas from rising and adds to the burden on your air-conditioning system. To overcome this heat, your air conditioner will have to operate almost constantly during summer weather (Fig. 13-2). Superheated attic air will actually scorch rafter boards and sheathing, wilt insulation, and cause shingles to cup and buckle.

Many newer homes—built with insulation, weather-stripping, doors, and windows that are designed to reduce energy losses—are more susceptible to condensation problems than older homes that have not been retrofited. During winter, an attic must have more ventilation than during summer because windows and doors are usually kept closed in cold weather.

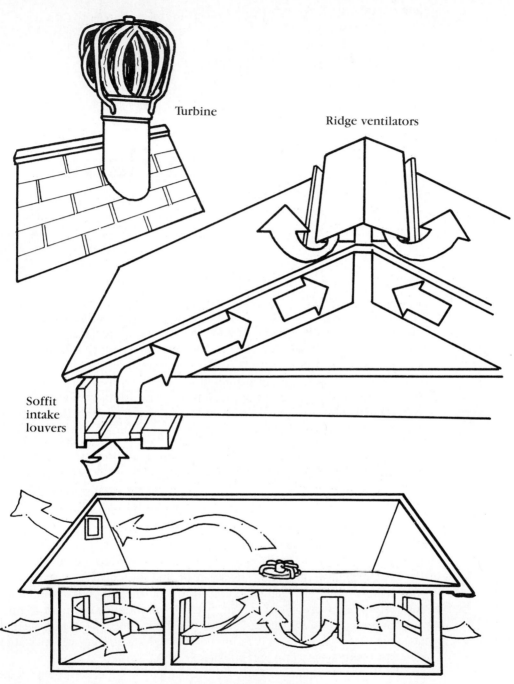

Turbine

Ridge ventilators

Soffit
intake
louvers

Cool air circulated by
a whole-house fan

Fig. 13-1 A balanced ventilation system.

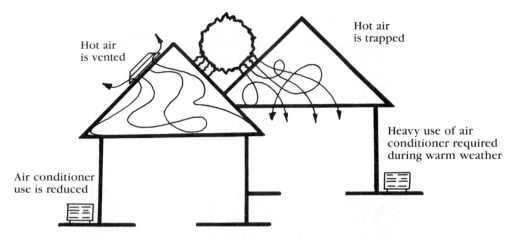

Hot air
is vented

Hot air
is trapped

Heavy use of air
conditioner required
during warm weather

Air conditioner
use is reduced

Fig. 13-2   Without adequate attic insulation, your air conditioner will work continuously during summer weather.

Unless water vapor—produced by using bathtubs, showers, and home appliances—that passes through the ceiling and collects in the attic is removed by adequate ventilation, it will soak and destroy insulation and perhaps even rot rafters.

Ice Dam Formation

In some areas of the country, ice dams form on the roof edges of poorly ventilated homes (Fig. 13-3). An *ice dam* will form when the heat from trapped attic air slowly melts snow on the middle and upper portions of a roof section. During very cold but sunny days, radiant heat will increase the flow of meltwater. As the water travels toward the edges of the roof, the meltwater cools because there is less heat or no heat from the attic, and there is more snow at the eaves. Some of the water will freeze and create a dam of ice.

As the meltwater increases, so does the accumulation of ice that eventually forces additional meltwater under the shingles and through the sheathing and into an attic or a wall. An ice dam can create enough force to lift some shingles, but most of the damage will not be found on the roof surface. The water will penetrate insulation materials and the ceiling. By the time you notice spots on the ceiling and peeling paint or plaster cracks, significant damage inside the home will have occurred.

Proper ventilation, along with at least an adequate amount of insulation, will prevent the uneven temperatures that cause ice dams. Because warm air rising through the attic or ceiling of the house is a key factor in ice dam formation, you can address the problem by reducing the heat flow by adding insulation and by increasing the number of *attic ventilators*. Additional cold air in the attic will decrease the amount of meltwater because the bottom of the roof will approach the outside temperature. The additional insulation will reduce the flow of heat rising from the living area.

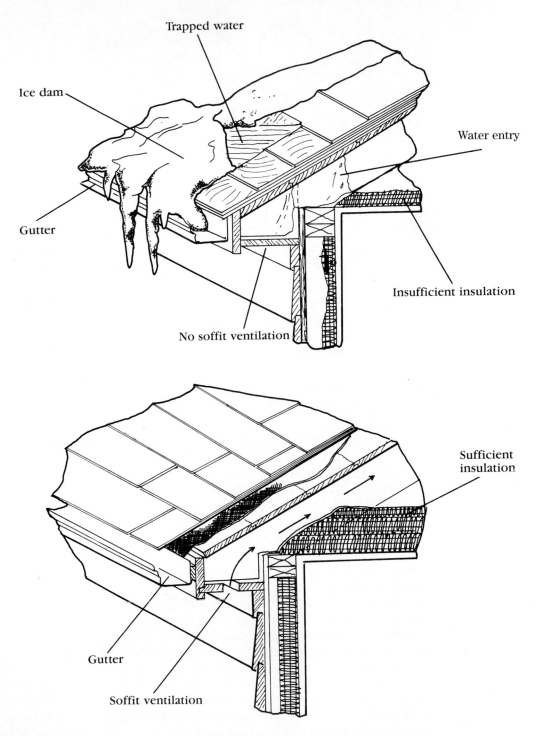

Fig. 13-3 Ice dam formation.

Alternative solutions to ice dam problems include treating the symptoms by installing a continuous sheet-metal strip on top of the shingles at the eaves. On steeply sloped roofs, the slick metal surface and gravity combine to discourage the formation of ice. You can also install heat cables that can be clipped at a zigzag angle to the eaves shingles, but this seldom solves the symptoms. In theory, the cables can be extended into the gutters to ward off freeze-ups.

Additional protection from ice and meltwater at the eaves can be obtained by installing a 3-foot-wide strip of roll roofing, selvage edge, or best of all, a water-and-ice-shield membrane on a new roof deck or where the old shingles have been torn from the surface. Several manufacturers offer rubberized asphalt and polyethylene sheets that bond directly to the roof deck and provide a backup barrier against meltwater from ice dams. Check with your local building supplier to see if the application of water-and-ice-shield membrane is recommended for your area. Also, keep in mind that this type of membrane makes an excellent underlayment for valleys or around skylights.

Types of Ventilators

Roof and *gable* ventilators allow superheated attic air to escape and prevent rain, snow, and insects from entering. Whole-house fans can be mounted in attic floors or in walls.

Roof-line louvers   Roof *louvers* (Fig. 13-4) are simple devices that cover the holes cut in the roof near the peak. The louvers allow hot air to escape as it rises.

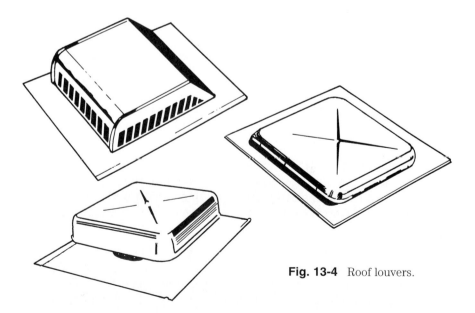

Fig. 13-4   Roof louvers.

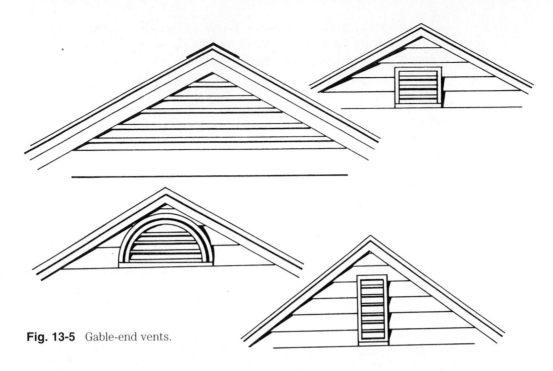

Fig. 13-5 Gable-end vents.

Gable-end vents   Gable vents (Fig. 13-5) are probably the most common and least useful type of ventilation unit. They are usually triangular or square and are flush or recess mounted.

Under-eaves vents   Under-eaves *soffit* ventilators (Fig. 13-6) are essential for a balanced ventilation system. About 2 square feet of soffit ventilation is needed for every 600 square feet of attic-floor area.

Wind-driven turbines   A turbine vent (Fig. 13-7) is designed to be driven by the wind from any direction. As the turbine spins, reduced air pressure in the throat of the unit draws hot, humid air from the attic.

Cupolas   *Cupolas* (Fig. 13-8) are essentially disguised large, wind-driven ridge ventilators. With a weather vane, a cupola can make an attractive roof structure.

Ridge ventilators   A ridge vent (Figs. 13-9 and 13-15) uses the natural flow of rising hot air from the eaves to the top of the ridge to exhaust superheated air from the attic. When combined with under-eaves vents or a whole-house fan, ridge vents provide a completely balanced ventilation system.

Whole-house fan   A *whole-house fan* (Fig. 13-10) is designed to help cool a house by drawing cooler air throughout the house during early morning and evening hours. A whole-house fan system combined with proper attic ventilation can reduce air-conditioning costs by 60 percent or more.

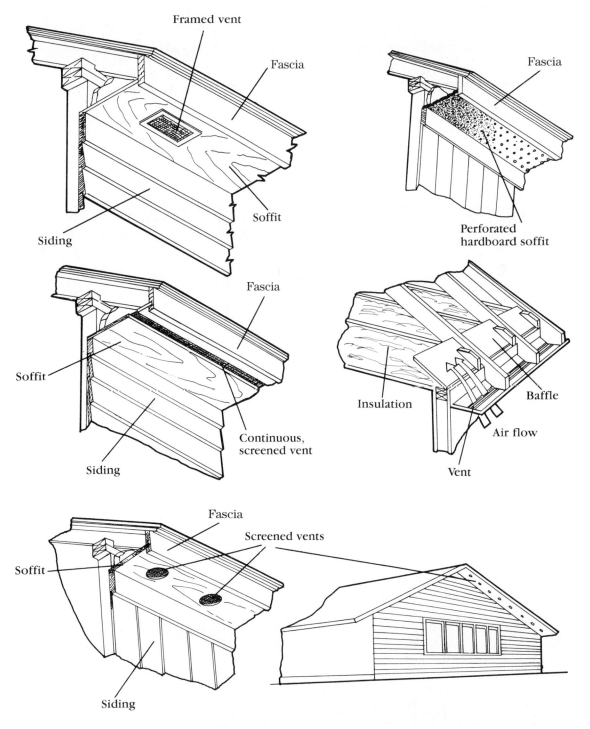

Fig. 13-6 Under-eaves vents.

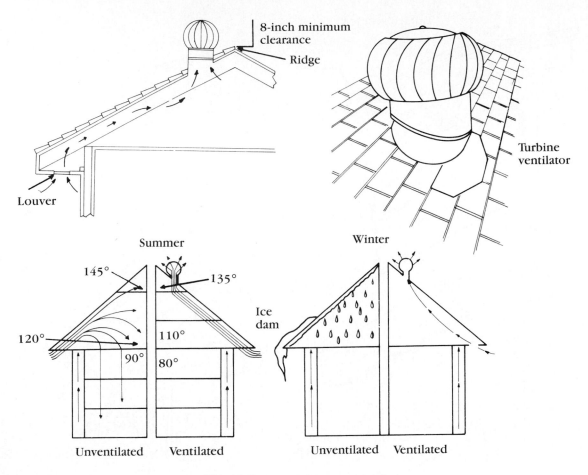

8-inch minimum clearance

Ridge

Louver

Turbine ventilator

Summer

Winter

145°

135°

Ice dam

120°

110°

90°

80°

Unventilated   Ventilated

Unventilated   Ventilated

Fig. 13-7 A wind-driven turbine vent.

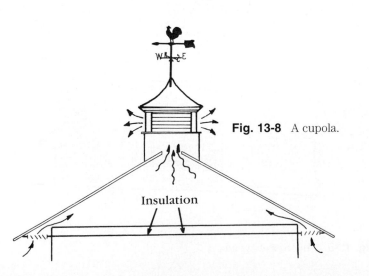

W   E

Fig. 13-8 A cupola.

Insulation

Fig. 13-9 A ridge ventilator.

Built-in baffle

Two connectors required per joint

End cap

Units are 10 feet long, and can be cut at 2-inch intervals as necessary

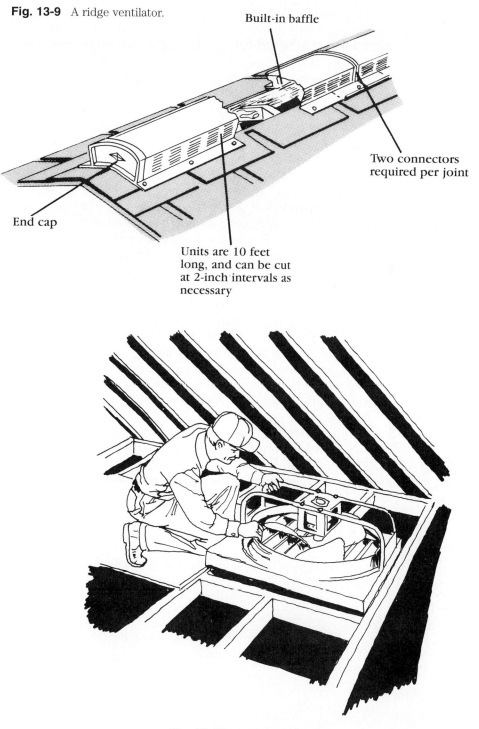

Fig. 13-10 A whole-house fan.

A cooler roof will last longer than one where the shingles are continually subjected to the expansion and contraction caused by the constant superheating and cooling of attic air throughout summer. To vent superheated attic air (which can reach 50 degrees higher than the outside temperature), install turbines—one 12-inch unit for about every 600 square feet of attic floor—near the peak of the roof or ridge vents mounted along the full length of the peak. Properly installed turbines or ridge vents will keep an attic within 5 to 10 degrees of the outside air temperature.

In temperate climates, turbines or ridge vents in combination with soffit vents can be used with a whole-house fan to virtually eliminate, except during the hottest summer weather, the need for air-conditioning. Whole-house fans must be installed in the attic floor in order for the fan to draw air directly from the living areas below. A whole-house fan can be installed in a wall but might not operate as efficiently as a floor-mounted unit.

A whole-house fan system is designed to be operated when outside temperatures drop to comfortable levels. Cool air is drawn through the lower-level windows of the house, and the fan pushes the air through the attic and out of the vents.

A whole-house fan is rated by its cubic-feet-per-minute (cfm) capacity to move air through a house. Properly installed, a whole-house fan system can evacuate the air in a typical house in 2 minutes. You can calculate the size of the fan you will need to cool your home by totaling the gross square footage of your home's living areas and multiplying by 3; multiply by 4 in very warm, humid areas. Add 15 percent if your home has dark-colored shingles. Do not include closet, stairwell, basement, and attic spaces in your living-space total. As an example for a single-story home that measures 40×50 feet (or 2000 square feet), multiplying by 3 equals 6000 square feet. Therefore, a whole-house fan with a minimum actual air delivery of 6000 cfm is needed to adequately ventilate the home. The fan's cpm rating should always be measured at 0.1 inch static pressure and, to reflect actual air delivery, the manufacturer's recommended louver must be in place.

Installing Vents

To install a vent, you'll need basic shingling tools as well as several carpentry tools. Have the following items on hand before you begin work:

- ladder
- roofer's hatchet
- tape measure
- chalk line and chalk
- nail pouch
- roofing nails
- nail bar

- utility knife
- roofing cement
- trowel
- screwdriver
- carpenter's level
- power saw, keyhole saw, or saber saw
- drill
- crayon or felt-tip marker
- 3-inch spike

Turbines

To install wind-driven turbine ventilators, first determine the number of 12-inch ventilators you'll need and note their locations. Along with the turbines, you'll need 2 square feet of soffit or eaves-louver area ventilation for every 600 square feet of attic area to allow sufficient outside air to enter the attic.

The best place to install a turbine is on the rear slope of the house about 8 inches from the ridge line and between roof rafters. Drive a 3-inch spike from the inside of the attic through the sheathing and the roof materials. On the roof, use the spike as a center point. With a felt-tip marker or crayon, lightly mark a circle that is wide enough to accommodate the throat of the turbine on the shingles. Carefully loosen, remove, and save the shingles around the circle. Remove as few shingles as is practical.

Again using the spike as a center point, mark a circle on the felt wide enough to accommodate the throat of the turbine. Remove the spike, cut through the sheathing, install the turbine, and replace and trim the shingles. The turbine must be watertight. Refer to the section on installing vent flanges in chapter 7.

The techniques for shingling around turbines are nearly identical to those for shingling around other roof obstacles.

Power Ventilators

Powered ventilators (Fig. 13-11) are installed essentially the same way as wind-driven turbines. Depending on the model you select, provisions must be made for thermostat, humidistat, and electrical hookups. Powered ventilators are typically factory-set to automatically switch on at 100 degrees and automatically turn off at 85 degrees.

Ridge Ventilators

To install ridge vents, first measure along the ridge to determine the number of vents required. Be certain to allow 6 inches at each end of the roof that is not to be cut. In other words, if the ridge measures 35 feet, the total length of ridge vents required is 34 feet. If a ridge vent must be cut to fit the ridge, make a cut where it will not damage the internal baffle.

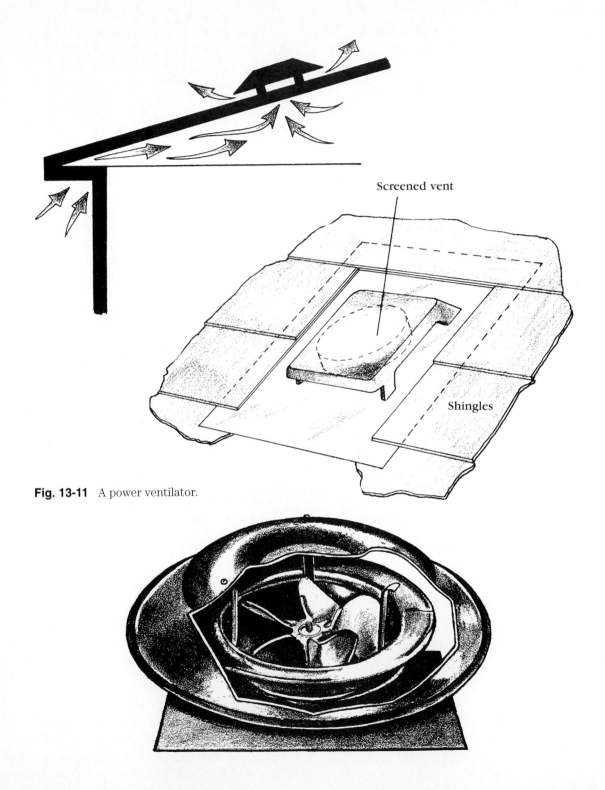

Screened vent

Shingles

Fig. 13-11 A power ventilator.

Use a pry bar (Fig. 13-12), claw hammer, or tear-off shovel to cut through the cap shingles and felt. Use a power saw (Fig. 13-13) to make a vent slot to within about 6 inches of the end of the ridge (Fig. 13-14) but be careful not to cut structural rafters or ridgepoles. The vent slot should be 2 to 3 inches wide for truss construction or 3 inches wide for ridgepole construction. Because instructions will vary somewhat with different ventilator products, be certain to follow the installation details for the product you purchase.

To align the vents, snap chalk lines on both sides of the vent-slot opening in the ridge. Center and install the 10-foot-long vents (Fig. 13-15) the entire length of the ridge so that the vent ends are flush with both rakes. Nail approximately every 12 inches and at each overlap of joints. Apply caulk if recommended by the manufacturer of the ridge-vent unit you are using.

Skylights

When properly sized and oriented, skylights (Figs. 13-16 and 13-17) can reduce energy consumption throughout the year. Recent improvements in the design of skylights include acrylic-domed units insulated with dead air space, self-flashing models that are fastened directly to the roof deck without a curb, flexible tubular models that are economical and easy to install between rafters, motor-driven vent windows, and models with shades or blinds.

During the summer, electrical lighting costs can be reduced significantly with the use of skylights. Some 90 percent of the electrical energy used for artificial lighting produces heat instead of light. The solar heat gain from skylights will be more than offset by the reduction of heat from decreased use of artificial lighting.

When you are planning the installation of a skylight, keep in mind the type of room in which the skylight is to be installed and how much direct sunlight will enter during each season. A skylight without a suitable interior cover or shade can contribute excessive heat gain. Shafted skylights usually require a mechanical or electrical retracting mechanism to employ a shade system.

Too much direct sunlight might even bleach carpet, furniture, or walls. One solution to receiving too much direct sunlight is to choose a tinted acrylic skylight. The white, bronze, or gray glazing helps block strong sunlight. The 3M Construction Markets Division offers aftermarket Scotchtint films, that are installed by professional applicators, for window and skylight energy control.

An alternative system for bringing natural light to an otherwise dark room is manufactured by Solatube. Solatube employees a 10-inch-diameter, roof-mounted dome and reflector, flashing, angle adaptors, extension tubes, mirrored tubing, and a prismatic light diffuser to evenly spread light. The tubing telescopes to fit between rafters and joists. Do-

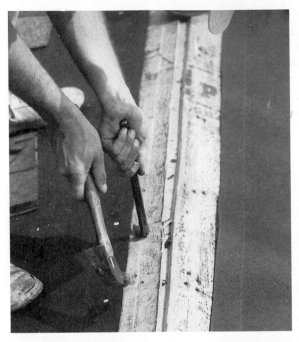

Fig. 13-12 Use hand tools to remove shingle caps, felt, and nails from the ridge.

Fig. 13-13 Use a power saw to cut a slot in the ridge center.

Fig. 13-14 Cut the vent slot about $1\frac{1}{2}$ inches on each side of the ridge peak.

Fig. 13-15 Position and snap together the 10-foot sections. Nail at precut holes.

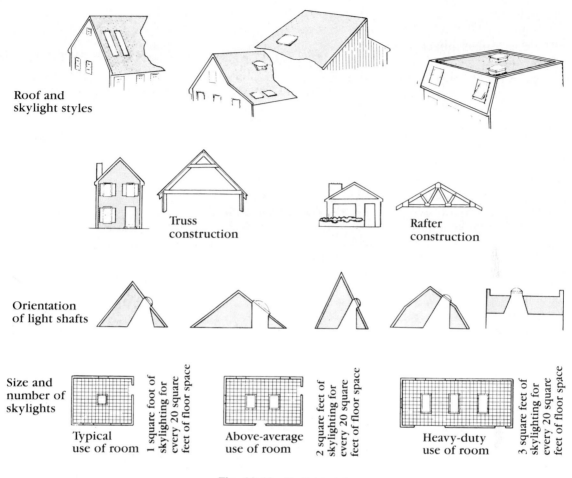

Roof and skylight styles

Truss construction

Rafter construction

Orientation of light shafts

Size and number of skylights

Typical use of room

1 square foot of skylighting for every 20 square feet of floor space

Above-average use of room

2 square feet of skylighting for every 20 square feet of floor space

Heavy-duty use of room

3 square feet of skylighting for every 20 square feet of floor space

Fig. 13-16 Skylight styles.

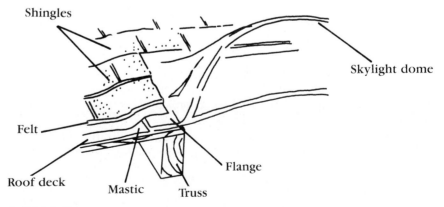

Shingles

Skylight dome

Felt

Flange

Roof deck

Mastic

Truss

Fig. 13-17 Self-flashing skylight design.

it-yourself kits are available or contractors will install the Solatube for about $300.

The Sun Tunnel offers do-it-yourself kits for two of its models that require about two hours of installation time. Both 16-inch and 20-inch Sun Tunnel models feature tubing that can flex around attic obstructions and ducts to eliminate framing, drywall, and repainting costs. You can install a series of vent tabs to create 78 inches of ventable air space. If ventilation is not needed, solid tabs can be installed.

Installation Guidelines

Skylights vary considerably, and each model has specific installation instructions. As with roof vents, the ideal time to install a skylight or a roof window is when you are shingling the roof. The tools and materials you'll need to install a skylight include:

- ladder
- roofer's hatchet
- nail pouch
- roofing nails
- nail bar
- tape measure
- chalk line and chalk
- utility knife
- roofing cement
- trowel
- carpenter's square
- power saw or hand saw
- keyhole saw
- drill
- lumber for headers and for the curb if it is not part of the sky light unit
- plasterboard, insulation, etc., if a light shaft is required.

As you plan your work, check the weather forecast and make certain you have allowed sufficient time to complete the project. Before you are ready to position the skylight, arrange for someone to help you carry the unit to the roof.

Installing a skylight can be only a little more complicated than adding roof ventilators or it can become a major renovation. See Figs. 13-18 through 13-20.

The size of the skylight opening in the roof is a good deal larger than for vents. The minimum skylight or roof-window area should be 10 percent of the floor space. Another important difference in installing a skylight is

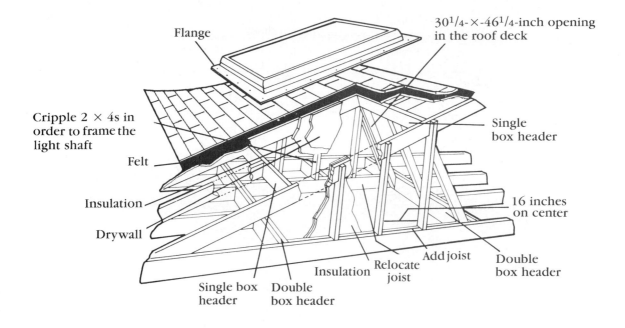

Fig. 13-18 Skylight installation details.

the additional flashing work required. With many models, you'll have to construct a curb from 2 × 4s; other units have their own integrated curb and flashing.

The Daylighter LongLite skylight by APC Corporation is an example of perhaps the easiest type of roof window to install because it can be placed directly between existing roof rafters or trusses and a minimum of framing is needed.

The installation position is marked by driving four spikes—one for each corner—from inside the attic through the sheathing and the roof materials. On the roof, use the spikes as guidelines to outline with a felt-tip marker or a crayon the area to be opened. Carefully loosen, remove, and save the shingles around the rectangle. Remove as few shingles as is practical.

Again using the spikes as guidelines, mark a rectangle on the felt. Remove the spikes and cut through the sheathing. At each end of the opening, install a cross brace between the rafters to form a rectangular frame.

Fasten the skylight to the wood framing with clips. A flat flange is part of the unit, and nail strips of aluminum flashing where the flange and roof join. Apply roofing cement as directed by the manufacturer as you trim and replace the surrounding shingles.

Skylights are flashed and shingled just like any other roof obstacle (see chapter 7). The key to a properly installed skylight is the flashing. If a skylight leaks, it is usually a result of improper installation where the shingles

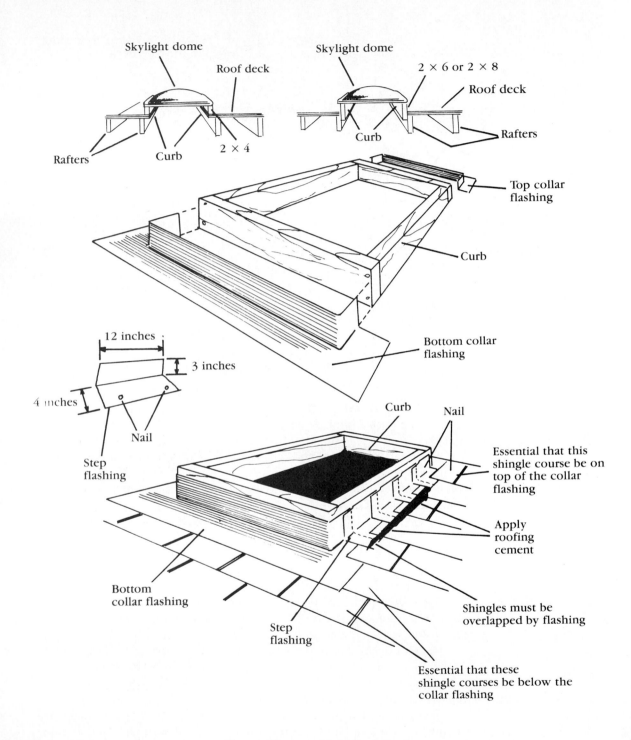

Fig. 13-19 Skylight flashing details.

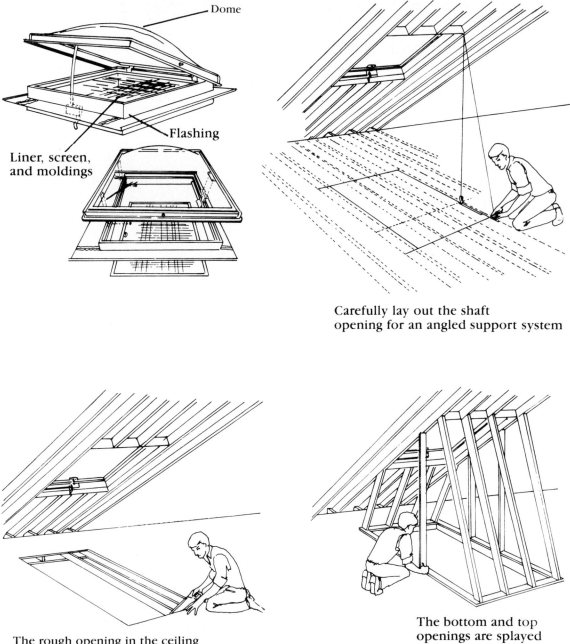

Carefully lay out the shaft
opening for an angled support system

The rough opening in the ceiling
will require insulation on four sides

The bottom and top
openings are splayed

Fig. 13-20 Installing a shafted skylight.

and the flashing join. At the "top" of the skylight, the intersecting course of shingles must rest snugly against, and be on top of, the aluminum or galvanized sheet-metal collar flashing. Use L-shaped aluminum or galvanized metal step flashing at the sides of the skylight, and weave the shingles with the step flashing. At the "bottom" of the skylight, the collar flashing must rest on top of one or two courses of shingles (depending on exactly where the skylight meets the shingle courses).

Chapter **14**

Repairs

A particularly harsh winter, an unusually persistent, heavy rainy season, or just one intense wind storm can result in isolated damage to your roof that needs immediate attention. If your roof is somewhere between 15 to 20 years old, seasonal or storm damage can reveal enough overall wear to indicate that you should consider replacement of the entire roof. For roofs only a few years old, missing shingle tabs, dislodged flashing, or limited damage from hail or a broken tree branch should require only spot repairs.

If your entire roof must soon be replaced, you still might want to postpone the work for a few weeks or months while you select and await delivery of materials, arrange financing, or while you determine whether or not you want to do the work yourself. If the weather has damaged many roofs in your area, it might take considerable time for you to receive several contractors' competitive estimates. See chapter 3 for guidelines on choosing a contractor.

Meanwhile, you might have to place plastic tarp around the base of a chimney or skylight or use fiberglass mesh and roof cement to temporarily repair flashing around vents, walls, and other intrusions in the roof surface. Individual asphalt-fiberglass shingles that are missing a few tabs can be replaced using the techniques in this chapter and by using the steps described in detail in chapter 5.

If strong winds have loosened, curled, or torn a few tabs or shingles, wait until the sun has warmed the damaged area and then use plastic-based roofing cement to secure the underside of the repaired tabs. For more seriously damaged areas, an effective but unattractive temporary repair to torn shingles can be made by driving nails through the exposed surfaces on both sides of the torn tabs, and then applying plastic-based roofing cement over the nailheads and the tears in the shingles.

Individual tabs, full shingles, or a series of full shingles that have been damaged can be removed and new tabs or full shingles can be slipped in as replacements. When you are replacing tabs or full shingles, keep in mind that each shingle is secured by two rows of fasteners providing double nailing and that most shingles are now manufactured with a self-sealing strip just above the exposure line. You will have to break the self-sealing bond and pry out the eight nails or staples: four fasteners from the undamaged course above and four fasteners holding each damaged shingle.

Use a putty knife or trowel to break the seal on the damaged shingles and gently break the seal on the course above the shingles to be replaced. Lift the tabs of the undamaged course above, and remove the fasteners at the watermarks from this course with pry bars; these fasteners hold the tops of the damaged shingles. Remove the fasteners located between the self-sealing strips and the watermark of the damaged shingles. Pull the damaged shingles free.

Position the new tabs or full shingles so that the replacement shingles align with the intact courses. Nail the replacements as well as the course above the area you replaced, and apply a few dabs of roofing cement to seal the repaired courses. The technique for replacing tabs and full shingles closely follows the storm-damage repairs shown in Figures 14-2 through 14-20.

The source or sources of roof leaks that cause water stains to appear on ceilings or walls are often far from where you see the damage. When searching for the source of leaks, take into account how much water is evident inside the house and how long it takes for interior moisture to appear after it begins to rain. If moisture appears on the ceiling or wall very soon after it starts to rain, check the flashing at the nearest intrusions—such as at vents, in a valley, along dormer walls, or around chimney flashings—on the roof surface.

The longer it takes for fresh moisture to appear—for example, on an inside ceiling—carefully look on the roof surface for suspected leaks at missing shingle tabs and worn flashing farther away from the location of the damaged interior. Remember that water can travel considerable distances along interior walls and drywall before the moisture becomes visible inside your home.

Repairing Storm Damage

Fixing shingle roof damage that results from wind, rain, hail or an ice storm is a typical roof repair project. When damage occurs to a section of roofing that has undamaged shingles above and below the shingles that are to be replaced, you'll have to lift the tabs of the undamaged course above the damaged shingles. Work carefully because each course of shingles is double-nailed and the seal-down strips inevitably will have been activated by the heat of the sun.

Figure 14-1 shows a broken tree limb—felled by a heavy ice storm—that became embedded like a spear in the ridge line of a home with five-year-old asphalt-fiberglass shingles. To begin making repairs on a project such as this, use a pry bar to pull the nails securing the first few damaged shingles (Fig. 14-2) and a utility knife, and if you have one, a tear-off shovel to cut and pull away additional shingles or courses (Figs. 14-3 and 14-4). Continue to pull nails until you can determine the extent of the damage and find out how many courses will have to be replaced (Fig. 14-5).

Because of the damage caused by the falling tree branch shown in Figs. 14-6 and 14-7, shingle courses must be removed from both sides of the roof ridge line. A circular saw can be used to remove a small portion of decking between trusses (Figs. 14-8 through 14-10). The damaged area of the roof deck is repaired by installing a new section of plywood (Figs. 14-11 and 14-12), new felt (Fig. 14-13), new shingles (Figs. 14-14 through 14-17), and new ridge capping (Figs. 14-18 through 14-20).

Repairs at the Eaves

If shingles at the eaves were properly installed, there is a 1-inch overhang that allows rainwater shed by the roof to flow directly into the gutters. When the exterior of an older building is remodeled, changes to the trim

Fig. 14-1 A heavy ice storm caused a broken tree limb to penetrate the ridge capping and the roof deck.

Fig. 14-2 Use a pry bar to begin removing damaged shingles.

Fig. 14-3 Use a utility knife to cut away damaged shingles.

Fig. 14-4 Remove additional shingles with a tear-off shovel.

Fig. 14-5 Remove nails and shingles until the damage to the roof deck is revealed.

Fig. 14-6 Pry damaged shingles from both sides of the ridge.

Fig. 14-7 Remove capping, shingles, and felt to uncover the damaged portion of the roof deck.

Fig. 14-8 Replace damaged roof decking with plywood of similar thickness.

Fig. 14-9 Make sure you have removed nails before you use a power saw to cut out the section of damaged plywood.

Fig. 14-10 Remove the damaged plywood from between trusses.

Fig. 14-11 Install the plywood patch.

Fig. 14-12 Trim felt to cover the patched plywood.

Fig. 14-13 Install felt to fit the areas to be reshingled.

Fig. 14-14 Replace shingles in the roof pattern by adding full or partial shingles.

Fig. 14-15 Continue to add shingles to the pattern. Be sure to install four nails for each full shingle.

Fig. 14-16 Lift tabs off the old shingle courses where necessary to maintain the shingle pattern of the repair courses.

Fig. 14-17 At the ridge, lap the top course of shingles to provide additional waterproofing.

Fig. 14-18   Install diamond-shaped caps to cover the ridge line.

Fig. 14-19   Install caps with a 6-inch overlap. Drive one nail on each side of the ridge line.

Fig. 14-20 Use a utility knife or tin snips to cut the last cap to fit.

and soffit or replacing gutters sometimes can result in problems at the eaves. Similar problems might result from the faulty installation of the eaves drip edge, the eaves border shingles and the first course of shingles, or a match-up of the eaves border shingles and the first course of shingles. Refer to the chapter 4 section on installing drip edge and the associated application tip for additional information on the proper installation of drip edge and border shingles at the eaves.

As shown in Fig. 14-21, the eaves border-shingle course and the first course of asphalt-fiberglass shingles are installed flush with drip edge at the eaves. Therefore, rainwater and meltwater has a tendency to fall short of the gutter. Combined with a low-slope roof, the resulting spillover will discolor the exterior wall and could potentially damage wall interiors if ice dams develop.

To correct an inadequate overhang at the eaves, first use a utility knife or tin snips to cut 6-inch-wide strips from the tops of individual shingles (Fig. 14-22) or from roll roofing. Use a flexible roofing material so that—after it is installed—the 1-inch overhang curls slightly toward the gutter. Because of the standard nailing pattern securing the above courses, remember that a strip wider than 6 inches will not fit under the bottom course of shingles at the eaves.

At the corner of the eaves and rake, lift up the first course of shingles and slide the first repair strip between the shingle and the upside-down border shingle (Figs. 14-23 through 14-25). This work should be easy

Fig. 14-21 The eaves-border shingles and the first course of shingles were installed to flush with the drip edge. Therefore, runoff falls short of the gutter.

Fig. 14-22 Use a utility knife or tin snips to cut repair strips from full shingles. Strips of roll roofing also can be used for repairs.

Fig. 14-23 Lift the bottom course and slide the repair strip between the border shingle and the first-course tabs.

Fig. 14-24 Lift the bottom course tabs to secure the repair strips with nails. Add a light coat of roofing cement to enhance waterproofing.

Fig. 14-25 Curl the bottom of the repair strips so that runoff from the roof will spill into the gutter.

because the bottom-course shingle tabs are not nailed when installed using the standard shingling pattern.

Use four roofing nails to secure each repair strip. Be certain to avoid a match-up at the watermarks by offsetting the repair strips and the bottom course by 6 inches. If the repair strips and the bottom course are allowed to match, leaks can result.

Roof Deck Repairs

If worn shingles must be removed from a roof, it is quite common to uncover areas on the roof *deck* where several feet of damaged plywood or damaged roof planking (Fig. 4-26) must be repaired or replaced before new roofing material is installed. A tape measure, a pencil or a marker, a pry bar, a claw hammer, and a circular saw should be all the tools you'll need to repair the decking. The replacement plywood or planking must be the same thickness as the original wood.

While no two repairs to a wood deck are identical, the general idea is to always cut away the damaged wood between trusses and insert new plywood or planking. Some undamaged wood might have to be cut away along with the broken planking or plywood to provide an adequate nailing base and truss support for the new wood. See Figs. 14-27 through 14-32.

Sometimes it is necessary to temporarily remove a section of plywood (Fig. 14-33) or planking (Figs. 14-34 and 14-35) during tear-off work. If the original roofing material was properly installed so that it extends underneath an overlapping deck section, using a pry bar is the best way to pull the nails and wood and access the remaining roofing material and debris to be torn out. Once the underlying roof section has been covered with new roofing material, the plywood or planking can be renailed (Fig. 14-36).

If broken *sheathing* or planking results from small cracks or at large knotholes a few inches wide, cut away the broken or rotted wood and cover holes with a patch of galvanized metal. Do not use metal patches on holes larger than 4-x-4 inches.

Moss and Algae

The appearance of moss, algae, mold, or mildew growing on shaded sections of your shingled roof could be a very minor cosmetic distraction from the overall attractiveness of your home or the warning sign of a much more serious problem.

The most likely cause for unsightly fungus on a roof in a high-moisture climate is lack of sunlight and reduced aeration due to continuous shade from tree branches or other tall shrubbery. Also, check the roof surface, gutters, and downspouts for debris such as leaves and twigs that might obstruct or dam a low-slope roof surface. Removing overhanging branches and gutter obstructions will often end unsightly fungus problems.

Because fungus requires persistent moisture to survive, fungus growth on your roof can be a sign that your shingles are retaining some moisture instead of draining all runoff. Where moss or fungus is prevalent, carefully check to determine if the roof deck seems weaker than other areas of the roof. If, when you walk on it, the deck sags, gives, or is spongy where the fungus appears, moisture damage to the deck from water penetration is a distinct possibility. The fungus itself will not damage asphalt-fiberglass based shingles.

If the deck is sound, compare the condition of shingles on other sections of the roof. If the wear on the shingles on all areas of the roof appears to be identical, try removing overhanging shrubbery or tree branches to eliminate the fungus. If the shingles with moss or fungus appear to be more aged, more brittle, or more worn than other shingles, repairing only the section with the fungus might be warranted.

While the best way to prevent recurring fungi is to promote airflow and sunlight by clearing a portion of the shrubbery shading your roof, you could remove the fungus by spraying the area with a mixture of detergent or bleach and pressurized water. Use four parts water to one part household bleach. In the process, however, you could dislodge a significant amount of surface granules and, therefore, lessen the expected life span of your shingles. Using a stiff brush or stiff broom or a wire brush also could damage the surface of the shingles.

In sections of the country where roof fungus is prevalent, consider an application of Safer's Moss & Algae Killer or the 3M Algae Block Copper Roofing Granule System. In the Northwest region of the United States, 3M Algae block is available as a special order on new orders of Pabco roofing shingles.

If you prefer not to trim your trees or shrubbery, consider installing a 97-percent-zinc strip of metal across the ridge line and at any point across the roof—such as dormers, vents, skylights, and chimneys—that interrupts the flow of rainwater down the roof surface. When it rains or when snow and ice melt, the zinc will gradually wash down the roof, but not "rust" or "bleed" from the strips or discolor the roof. The small amounts of disbursed zinc will inhibit the growth of moss and fungus on the surfaces of shingles. A $2^1/_2$-inch-x-50-foot zinc strip lists for about \$27 in the ABC Supply Company catalog.

Fig. 14-26 Broken, damaged, or rotted planking or plywood is commonly found when worn roofing is removed from a roof deck.

Fig. 14-27 Replace damaged planking or plywood with new wood that is the same thickness as the original deck material.

Fig. 14-28 Use a circular saw to cut away the damaged wood between trusses.

Fig. 14-29 A pry bar will make easy work of removing nails securing the damaged wood.

Fig. 14-30 Break loose damaged wood.

Fig. 14-31 Remove enough planking so that you have an adequate base to nail the new patch between trusses.

Fig. 14-32   Nail the new wood in place.

Fig. 14-33   Use a pry bar to remove nails that hold worn flashing or debris.

Fig. 14-34   For some types of tear-off work, you'll have to remove the planking temporarily to permit access underneath an overlapping deck section.

Fig. 14-35   Use a nail bar and a claw hammer to pull nails.

Fig. 14-36 After the underlying roof section is foamed, renail the planking.

Appendix

Frequently Asked Questions

*T*his appendix contains the answers to home-owners' most frequently asked roofing questions. Where detailed answers and related issues or more information is available, concise answers are followed by suggestions to see specific chapter headings, subheadings, and illustrations found throughout the book.

Should I repair or replace my current asphalt-shingled roof? If your shingled roof is less than 15 years old and leaks are apparent, minor repairs will probably solve the problem. If the shingles on your home were installed 15 to 20 or more years ago, carefully inspect the condition of your roof, because the likelihood of reroofing increases with the age of the roofing materials. *See* chapter 1, Inspection Procedures, and chapter 14, Repairs.

How can I find a reliable roofing contractor in my area? Ask neighbors, friends, and relatives for the names of recommended roofing contractors. Any prospective contractor or remodeler should be willing to supply references from jobs underway or work recently completed. You can telephone the National Roofing Contractors Association at (800) USA-ROOF for a computerized referral service for your area's association members. *See* chapter 3, Choosing a Contractor; Sources, Associations, and Institutes.

Is roofing work a realistic do-it-yourself project? Most roofing work is well within the capabilities of many do-it-yourselfers. Nevertheless, some materials and roof work should be done by professionals. Hire a contractor to work on slate, copper, steep-sloped roofs, roofs that are several stories above ground, hot-tar-and-gravel *build-up roofing*, foam-roofing products, or single-ply (usually synthetic rubber) membranes (such as *EPDM*, ethyl-ene propylene diene momomer or *CSPE*, chlorosulphonated polyethylene).

If you can afford to purchase copper or slate for your roof, you can afford to pay someone to install the materials. The application of tar-and-gravel roofs is dangerous work and it requires very expensive specialized equipment. Foam roofing and nearly all single-ply roofing products (EPDM and CSPE) cannot be warranted unless contractors trained and licensed by the product manufacturer do the work.

If you decide to reroof yourself, first become an educated consumer who is prepared to talk with potential suppliers and contractors. Study the sections of this book that describe the different types of roofing products commonly available and obtain information provided by the manufacturers.

Roofing your home could easily mean lifting several tons of roofing material. You will have to work at a pace that is compatible with your health and conditioning. During warm weather, you must take frequent breaks and drink a lot of water. Stay off the roof during very hot or very cold weather, when it rains, or when it is very windy. Don't attempt to walk on a roof that has frost, ice, or snow on it.

Examine each section of your roof to determine if the pitch is too steep or if you feel you will be too high off the ground to work confidently.

See chapter 1, Roof Conditions and Materials; chapter 3, Preparations; Roof-It-Yourself Safety Guidelines; chapter 5, Starting Points.

If neighbors are replacing roofs on their homes, is mine due for reroofing?   Don't be rushed into a hasty decision by the actions of neigh-bors. If you live in a development where all the houses were constructed within a year or so of each other and quite a few of your neighbors have been reroofing their homes, it is likely that your roof is ready for replace-ment. Salespeople for roofing contractors will often work an entire neigh-borhood, convincing potential customers that every roof in the area is worn out. *See* chapter 1, Roof Conditions and Materials.

Why are many homes with wood-shingled roofs being reroofed with other types of roofing materials?   Because of the inherent flammability of wood, some communities now restrict or ban the installation of wood roofing shingles and shakes. Homeowner insurance rates for wood roofs are higher and insurance might not be available for a home with a wood roof. Make certain your *local building code* permits the installation of wood shingles or shakes on roofs in your community. Keep in mind that local codes and regulations are often based on or adopted from the *Uniform Building Code*, which is published about every three years by the International Conference of Building Officials. Also, treating wood shingles and shakes with chemicals to resist fire adds about 25 percent to the cost of the material. *See* chapter 1, Wood; chapter 9, Wood Shingles and Shakes.

I have one layer of worn asphalt shingles on my home. Do I have to remove the old shingles?   Some salespeople recommend strongly that all old shingles on any home be torn off before a new layer of shingles is installed. Actually, the old asphalt shingles must be torn off only when they are buckled or warped or if the roof already has two layers of shingles. *See* chapter 1, Roof Conditions and Materials; chapter 4, Tearing Off and Drying In; Resources, Recycling.

If worn shingles must be removed from my roof, how much will it add to the cost of the job and what happens to the debris? As an approximation, you can expect the cost of removing the old roof to double your estimated expense for simply recovering or shingling over the top of the worn layer. Be sure to research local landfill restrictions and check local government and business agencies to see if your community has one of the few businesses that recycle roofing materials. For example, Reclaim Inc. of Tampa Bay, Florida, currently recycles worn asphalt shingles into street-paving and road-patching materials ([813] 935-8533). *See* chapter 1, Roof Conditions and Materials; chapter 4, Tearing Off and Drying In; Resources, Recycling.

How many layers of asphalt shingles can be installed over a worn layer of shingles? One additional layer of new shingles can be installed over a layer of worn—but not warped or buckled—shingles. Don't believe it if someone tells you that you can get away with three or more layers of shingles. Even if the top layer of shingles is not curled or warped, standard roofing nails will not adequately penetrate the roof deck through more than two layers of shingles. Also, many residential buildings are not designed to handle the additional stress caused by what would be several tons of roofing material. *See* chapter 1, Roof Conditions and Materials; chapter 4, Tearing Off and Drying In, Figs. 4-1 and 4-2.

What is the most popular roofing product for homeowners and contractors? Because of the comparatively low initial cost and ease of application, three-tab, 20-year, asphalt fiberglass-based shingles are by far the most common roofing material used on recently constructed or reroofed homes. About 80 percent of American homes are roofed with asphalt or asphalt-fiberglass shingles. *See* chapter 1, Shingle Weights and Life Spans, Fig. 1-1.

Are all asphalt shingles three-tab, 12-×-36-inch strips? Asphalt-fiberglass shingles are produced in many brand names, colors, styles, textures, thicknesses, dimensions (including metric), and weights. Some manufacturers offer asphalt fiberglass-based strip shingles designed with a layered texture that resembles wood shakes, shingles with two cutouts, or strip shingles with no cutouts. Some diamond, hexagonal, and T-shaped shingles are also available for high-wind areas. *See* Table 1-1 for a comparison of common styles, weights, and dimensions; chapter 1, Asphalt-and Fiberglass-Based Shingles.

My favorite local home-improvement supplier doesn't stock and isn't familiar with the wide variety of shingle weights available. What can I do to get more information? Keep in mind that you are making a major purchase. According to the Home Improvement Research Institute, U.S. homeowners spent $87.7 billion on home-improvement products in 1994. Be prepared to shop around at several stores and suppliers. Don't be surprised if some of your questions about materials go unanswered. If you can't find local distributors or wholesalers for the products you want, consult the Resources section in this book for the dealers nearest you. Many manufactures provide toll-free telephone numbers and will send you detailed product comparison information. *See* chapter 1, Roof Conditions and Materials; Resources.

What color shingle is best for my roof?   If summer cooling or winter heating is a very important factor in your location, select an appropriate light color (for the Sunbelt states) or a dark color (for the Snowbelt regions). Otherwise, pick a roof-shingle color that best suits the color scheme of the entire house. A balanced ventilation system is significantly more important to efficiently cooling and heating your home than is shingle color. *See* chapter 1, Shingle Color; chapter 13, Ventilation.

Why won't the color from the supplier's or salesperson's samples exactly match the look of the finished roof?   Samples are usually small pieces of roofing material seen at close range and indoors under artificial lighting. Your installed roof will appear in large sections, be subject to natural, outdoor lighting, and will be viewed from a distance. Also, product color batch runs vary slightly during the shingle manufacturing process.

Is it necessary to remove the cellophane strips from the backs of asphalt shingles when the shingles are applied to the roof?   There is no need to pull the cellophane strips from the backs of shingles. Avoid the time-consuming effort, the aggravation, and the unwarranted folklore of needlessly pulling off the adhesive protective tape on the back of each shingle. Manufacturers place a protective strip across the back of each shingle to keep the self-sealing feature on the fronts of the individual shingles from activating while the shingles remain in paperbound bundles. Once shingles are installed, the protective tape will no longer prevent sunlight from activating the adhesive sealant. If a few protective tapes come loose while you are breaking apart bundles, pull off whatever tape easily comes off and continue shingling. *See* chapter 5, Shortcuts and Back-Saving Tips.

Can the white lines printed on felt be used for shingle alignment guidelines when installing shingles?   The printed white lines on felt are fine for aligning overlapping courses of felt but cannot be used to align shingle courses. See chapter 4, Laying Felt; Fig. 4-22.

Is it necessary to install felt over a wood deck before applying asphalt-fiberglass shingles?   Felt must be installed over a wood deck as a vapor barrier between asphalt shingles and the wood. Without the felt layer, the wood planking or plywood deck—dried by superheated temperatures found in most attics—will draw moisture (including tar and oils from the asphalt) from shingles nailed directly to the wood deck. The result will be prematurely worn, cracked, or brittle shingles, as well as shingles stuck to the wood surface. *See* chapter 3, Estimating Materials (Felt); chapter 4, Laying Felt; chapter 5, Over-the-Top Shingling.

Is it necessary to install a layer of felt when you reroof over one layer of shingles? If you will not be removing the old shingles when you are applying a second layer of shingles, you do not need to install a layer of felt. *See* chapter 3, Estimating Materials (Felt); chapter 5, Over-the-Top Shingling.

What causes an ice dam to form on a roof? Inadequately ventilated soffits and attics cause ice dams to form when the heat from trapped attic air slowly melts snow on the middle and upper portions of a roof section. As the water travels toward the edges of the roof, the meltwater cools because there is less heat or no heat from the attic and there is more snow at the eaves. Some of the water freezes and creates a dam of ice.

As the meltwater increases, so does the accumulation of ice that eventually forces additional meltwater under the shingles and through the sheathing and into an attic or a wall.

Proper ventilation and adequate insulation will prevent the uneven temperatures that cause ice dams. You can address the problem of ice dams by reducing the heat flow by adding insulation and by increasing the number of attic ventilators. More cold air in the attic will decrease the amount of meltwater because the bottom of the roof will approach the outside temperature. The additional insulation will reduce the flow of heat rising from the living area. *See* chapter 13, Ice Dam Formation; Fig. 13-3.

Should I apply a self-adhering, water-and-ice-shield membrane directly to the wood deck or on top of the first course of felt? On a new roof deck or where the old shingles have been torn from the surface, additional protection from ice and meltwater at the eaves can be obtained by installing a 3-foot-wide strip of water-and-ice-shield membrane that will seal around the nails used to secure the shingles installed along the eaves.

Several manufacturers offer rubberized asphalt and polyethylene sheets that bond directly to the roof deck and provide a backup barrier against meltwater from ice dams. Some products advise that materials be applied directly to the wood deck. Many roofers choose to apply the felt and then the membrane so that the wood deck does not draw moisture from the membrane and prematurely age the water-and-ice-shield material. *See* chapter 3, Estimating Materials (Felt); chapter 4, Laying Felt; chapter 13, Ice Dam Formation.

When drip edge is applied at the eaves, should the felt go over or under the drip edge? Many professional roofers working with asphalt-fiberglass shingles install the strips of drip edge on top of the starter course of felt at the eaves (Fig. 5-31) to help keep the felt secure from wind gusts while work is in progress. Others insist that drip edge be installed along the eaves under the starter course of felt so that any rainwater able to penetrate overlapping courses of installed shingles will encounter the felt as a "back-up" moisture barrier and drain into the gutter. The suggestion is that such runoff otherwise could be trapped by the lip of the 3-inch-wide drip edge at the eaves if the roofer installs the drip edge on top of the layer of felt. In reality, a layer of felt under asphalt-fiberglass shingles provides little value as a "back-up" moisture barrier for the shingles because the felt has thousands of nail holes or staple holes in it caused by the installation of the felt itself and the overlapping courses of shingles.

If drip edge installed over felt at the eaves of an asphalt-shingle roof is trapping enough rainwater runoff to cause any sort of damage to your roof deck, your roof wore itself out a very, very long time ago.

The recommendation for placing drip edge underneath the felt at the eaves on all types of roofs apparently stems from the requirement for waterproofing the underlayment for clay and tile roofing materials. Many homeowners are not aware that clay and cement tiles are not intended to provide waterproof roof surfaces; rainwater is shed by the underlayment on such roofs. Furthermore, where local codes allow it, many builders and roofing contractors install only a minimum of 30-pound felt as a watertight underlayment for clay tile or cement tile roofs. Therefore, on tile roofs, the drip edge must be placed under the felt at the eaves. There seems to be widespread agreement that drip edge is correctly applied on top of the felt at the rakes (Fig. 5-32) for all types of roofing materials. *See* chapter 1, Other Materials; chapter 4, Laying Felt; Installing Drip Edge; Figs. 5-31 and 5-32.

For the starter course, is it acceptable to rotate a full shingle under the first course or is it necessary to cut about 5 inches from the shingle so that the self-sealing compound is at the bottom of this starter course? Many professional roofers prefer to apply a so-called upside down course by rotating full shingles in order to install eaves border shingles (Fig. 5-33). The alternative is to trim about 5 inches from the starter course shingles to ensure that the sealing strip is close to the edge of the overlapping (right-side-up) first course of shingles (Fig. 5-30, see label "starter course"). Application details vary with product manufacturers; some require a 12-inch-wide starter roll that has no self-adhesive strip as the eaves border material. Other manufacturers suggest trimming the upside-down course. And still other manufacturers allow a rotated full shingle. Experienced roofers know that each method is acceptable and none of the three methods will change the expected life span of the roof. *See* chapter 5, The 45 Pattern; Figs. 5-30 and 5-33.

Is applying six nails per shingle worthwhile in a location or climate where high winds are prevalent? The practice of *six nailing* instead of fastening each shingle with four nails was originally an option contractors sometimes specified and homeowners sometimes requested prior to the widespread advent of sunlight-activated, seal-down adhesive strips on asphalt-fiberglass shingles. Sunlight-activated adhesive strips provide a much stronger bond than the addition of two more nails. An additional two nails per lightweight asphalt shingle won't make much difference in hurricane gusts that approach 200 miles per hour.

Because of the extensive damage to roofs caused by 1992's Hurricane Andrew, in 1993 the Asphalt Roofing Association recommended nails as the preferred fastener for shingles. In 1994, the Southern Building Code Congress suggested using nails as the preferred fastener for shingles and using six nails in high-wind locations.

If constant high winds or potential hurricane winds are a feature of your area, a better solution than six nailing is to consider applying a heavier roofing material, such as asphalt/fiberglass architectural or so-called laminated shingles, on your home. An American Plywood Association report on 1992's Hurricane Iniki, which struck the Hawaiian island of Kauai, revealed that architectural shingles remained largely intact, "with minimal on no damage during the hurricane." *See* chapter 1, Other Materials; Resources.

Will adding a dab of roof cement under tabs help seal an asphalt-shingled roof? Sunlight-activated sealing strips makes it unnecessary to hand seal shingles with roof cement.

Is there a remedy for water stains on drywall caused by a leaky roof? A 50-50 mix of chlorine and water can sometimes remove water stains. In climates with low humidity, you probably will have to just scrub and repaint. Dab on some of the chlorine-and-water mixture and wipe, repeat, and wait to see if the water damage was only minor. Otherwise, more extensive repairs to the drywall will be needed.

Resources

The sources listed here can be contacted for product information or for locating product manufacturers or dealers nearest you. While efforts have been made to obtain the latest information, it is possible that changes have occurred since the list was compiled.

You can reach toll-free directory assistant by dialing (800) 555-1212 or by calling directory assistance for the city listed for the company or organization you want to reach. Additional telephone listings can be found by using CD-ROM products such as Direct Phone (8 million business listings) or Free Phone (lists all AT&T toll-free 800 numbers). Some libraries have similar CD-ROM products available.

At the reference desk of your local library, you can research products and companies in the *Thomas Register, Sweet's Catalogs, Hoover's Handbooks, Moody's Manuals, Ward's Business Directory* or the *U.S. Industrial Directory*. Information about associations can be found in the *Encyclopedia of Associations* or the *National Trade and Professional Associations of the U.S.*

Retail home centers usually have displays and product information guidance, how-to videos, and in-store clinics and do-it-yourself advice from staff. Some stores provide space for weekend or evening classes on popular topics. If you have difficulty locating any of the materials, tools, or equipment described in this book, contact the ABC Supply Company—the largest distributor of roofing supplies with 110 locations—for a free catalog. The address and toll-free number are listed under the Publications and Videocassettes subheading.

Associations and Institutes

AMERICAN BUILDING MATERIALS
Distributors Association
1417 Lake Cook Rd., #130
Deerfield, IL 60015
(708) 945-7201

AMERICAN CANVAS INSTITUTE
10 Beach St.
Berea, OH 44017

AMERICAN HOMEOWNERS FOUNDATION
(model homeowner/remodeler contract)
1724 S. Quincy St.
Arlington, VA 22204
(703) 979-4663

AMERICAN IRON AND STEEL INSTITUTE
1101 17th St. N.W.
Suite 1300
Washington, D.C. 200036
(800) 797-8335

AMERICAN RENTAL ASSOCIATION
1900 19th St.
Moline, IL 61265
(309) 764-2475

AMERICAN SUBCONTRACTORS ASSOCIATION
1004 Duke St.
Alexandria, VA 22314
(703) 684-3450

AMERICAN WOOD COUNCIL
1111 19th St., NW
Washington, DC 20036
(202) 463-2700

ASPHALT ROOFING MANUFACTURERS
ASSOCIATION
6000 Executive Blvd.
Suite 201
Rockville, MD 20852
(301) 231-9050

CANADIAN COPPER & BRASS DEVELOPMENT ASSOCIATION
10 Gateway Blvd., Suite 375
Don Mills, Ontario M3C 3A1
(416) 421-0788

CANADIAN HOME BUILDERS' ASSOCIATION
150 Laurier Avenue West, Suite 200
Ottawa, Ontario K1P 5J4
Canada

CALIFORNIA REDWOOD ASSOCIATION
405 Enfrente Dr., Suite 200
Novato, CA 94949
(415) 382-0662

CONVEYOR EQUIPMENT MANUFACTURERS
ASSOCIATION
932 Hungerford Dr., #36
Rockville, MD 20850
(301) 738-2448

COPPER DEVELOPMENT ASSOCIATION, INC.
260 Madison Ave.
New York, NY 100016
(212) 251-7200

HAND TOOLS INSTITUTE
707 Westchester Ave.
White Plains, NY 10604

HOME AIR COMPRESSOR ASSOCIATION
211 East Ontario, Suite 1300
Chicago, IL 60611

HOME IMPROVEMENT RESEARCH INSTITUTE
400 Knightsbridge Pkwy.
Lincolnshire, IL 60069
(708) 634-4368

HOME VENTILATION INSTITUTE
30 West University Dr.
Arlington Heights, IL 60004
(312) 394-0150

JAPAN PHOTOVOLTAIC ENERGY ASSOCIATION
c/o Kyocera Corp.
5-22 Kita-inouecho, Higashino Yamashina-ku
Kyoto 607
Japan
075-592-3861

METAL LADDER MANUFACTURERS ASSOCIATION
P.O. Box 580
Greenville, PA 16125
(412) 588-8600

NAHB NATIONAL RESEARCH CENTER
400 Prince George's Blvd.
Upper Marlboro, MD 20772
(301) 249-4000

NATIONAL ASSOCIATION OF THE REMODELING INDUSTRY
1901 North Moore St., Suite 808
Arlington, VA 22209
(800) 440-6274 or (703) 276-7600

NATIONAL ASSOCIATION OF HOME BUILDERS
15th and M Sts., NW
Washington, D.C. 20005
(800) 368-5242

NATIONAL PAINT & COATINGS ASSOCIATION
1500 Rhode Island Ave. NW
Washington, DC 20005
(202) 462-6272

NATIONAL ROOFING CONTRACTORS ASSOCIATION
8600 Bryn Mawr Ave.
Chicago, IL 60631
(800) USA-ROOF (800) 872-7663
for a computerized referral
service for local association members.

NATIONAL TILE ROOFING MANUFACTURERS
ASSOCIATION
3127 Los Feliz Blvd.
Los Angeles, CA 90039
(800) 248-8453

POWER TOOL INSTITUTE
P.O. Box 818
Yachats, OR 77498
(503) 547-3185

RED CEDAR SHINGLE & HANDSPLIT SHAKE BUREAU
Suite 275, 515 116th Ave. NE
Bellevue, WA 98004
(206) 453-1323

ROOFING INDUSTRY EDUCATIONAL INSTITUTE
14 Inverness Dr. E.,
Building H, Suite 110
Englewood, CO 80112-5608
(303) 790-7200

SCAFFOLD INDUSTRY ASSOCIATION
14039 Sherman Way
Van Nuys, CA 91405
(818) 782-2012

SINGLE-PLY ROOFING INSTITUTE
104 Wilmont Rd., Suite 201
Deerfield, IL 60015
(708) 940-8800

SMALL HOMES COUNCIL-BUILDING RESEARCH
COUNCIL
University of Illinois
One East St. Mary's Rd.
Champaign, IL 61820
(217) 333-1801 (Mon.–Thurs., 9:00 A.M.
to 12:00 P.M. only)

SOLAR ENERGY INDUSTRIES ASSOCIATION, INC.
77 N. Capitol St., NE, #805
Washington, D.C. 20002
(202) 408-0660

SPECIALTY TOOL & FASTENER DISTRIBUTORS
ASSOCIATION
500 Elm Grove Rd.
Elm Grove, WI 53122
(Phone) 784-4774

WESTERN RED CEDAR LUMBER ASSOCIATION
1200-555 Burrand St.
Vancouver, British Columbia V7X 1S7
Canada
(604) 684-0266

WESTERN WOOD PRODUCTS ASSOCIATION
Yeon Building
522 S.W. Fifth Ave.
Portland, OR 97204
(503) 224-3930

Asphalt/Fiberglass-Based, Architectural/Laminated Heavyweight Shingles

BIRD ROOFING DIVISION
(Architect Limited Editon, 370 pounds per square, 40-year warranty)
1077 Pleasant St.
Norwood, MA 02062
(800) 247-3462 or (551-0656)

BPCO ROOFING DIVISION
(Esgard Signature Series, 340 pounds per square, 35-year warranty)
850 La Fleur St.
LaSalle, Quebec H8R 3H9
Canada
(514) 364-0671

CELOTEX CORPORATION
Roofing Products Division
(Presidential Shake, 360 pounds per square, 40-year warranty)
P.O. Box 31602
Tampa, FL 33631
(800) 235-6839 or (813) 873-1700

CERTAINTEED CORPORATION
(Carriage House Shingles, lifetime, limited warranty)
P.O. Box 860
Valley Forge, PA 19482
(800) 441-6720, (800) 274-8530, or (215) 341-7000

ELK CORPORATION
(Prestique Plus, 360 pounds per square, 40-year warranty)
P.O. Box 500
Ennis, TX 75119
(800) 288-6789 or (214) 851-0400

GAF CORPORATION
(Timberline Ultra, 40-year warranty)
1361 Alps Rd.
Wayne, NJ 07470
(800) 766-3411 or (201) 628-3000

GEORGIA-PACIFIC CORPORATION
(Summit III, 300 ponds per square, 35-year warranty)
133 Peachtree St., NW
Atlanta, GA 30303
(800) 284-5347, (800) 423-2800, or (800) 765-5402

GS ROOFING PRODUCTS COMPANY
(High Sierra, 40-year warranty)
2900 Bird St.
Charleston Heights, SC 29405
(800) 777-2563

MANVILLE CORPORATION
(Woodlands Premier, 360 pounds per square, 40-year warranty)
P.O. Box 5108
Denver, CO 80217
(800) 654-3103

OWENS-CORNING ROOFING PRODUCTS
(Oakridge Shadow, 350 pounds per square, 40 year warranty)
Fiberglas Tower
Toledo, OH 43659
(800) 342-3745

TAMKO ROOFING PRODUCTS, INC.
(Heritage Premium Shadowtone Series,
40-year warranty)
P.O. Box 1404
Joplin, MO 64802
(800) 641-4691

Asphalt/Fiberglass-Based Shingles and Roll Roofing

BIRD, INC.
Bird Roofing Division
1077 Pleasant St.
Norwood, MA 02062
(800) 247-3462 or (551-0656)

BPCO ROOFING DIVISION
850 La Fleur St.
LaSalle, Quebec H8R 3H9
Canada
(514) 364-0671

CELOTEX CORPORATION
Roofing Products Division
P.O. Box 31602
Tampa, FL 33631
(800) 235-6839 or (813) 873-1700

CERTAINTEED CORPORATION
P.O. Box 860
Valley Forge, PA 19482
(800) 441-6720, (800) 274-8530,
or (215) 341-7000

CHAMPION INTERNATIONAL CORPORATION
Building Products Division
1 Champion Plaza
Stamford, CT 06921
(203) 358-7000

ELK CORPORATION
P.O. Box 500
Ennis, TX 75119
(800) 288-6789 or (214) 851-0400

FLINTKOTE
GS Roofing Products Company Inc.
5525 MacArthur Blvd.
Irving, TX 75038
(214) 580-5604

GAF CORPORATION
1361 Alps Rd.
Wayne, NJ 07470
(800) 766-3411 or (201) 628-3000

GEORGIA-PACIFIC CORPORATION
133 Peachtree St., NW
Atlanta, GA 30303
(800) 284-5347, (800) 423-2800,
or (800) 765-5402

GOODYEAR TIRE & RUBBER COMPANY
Roofing Systems
1144 E. Market St.
Akron, OH 44316
(800) 992-7663

W. R. GRACE & COMPANY
Construction Products Division
Ice & Water Shield Membrane
62 Whittlemore Ave.
Cambridge, MA 02140
(800) 558-7066

GS ROOFING PRODUCTS COMPANY
2900 Bird St.
Charleston Heights, SC 29405
(800) 777-2563

LESLIE-LOCKE, INC.
Customer Service Dept., Suite F6300
4501 Circle 75 Parkway
Atlanta, CA 30339
(800) 755-9392

TAMKO ROOFING PRODUCTS, INC.
P.O. Box 1404
Joplin, MO 64802
(800) 641-4691

UNITED STATES GYPSUM COMPANY
101 South Wacker Dr.
Chicago, IL 60606
(800) 347-1345

MANVILLE CORPORATION
P.O. Box 5108
Denver, CO 80217
(800) 654-3103

ONDULINE ROOFING PRODUCTS
4900 Onduline Dr.
Fredericksburg, VA 22401
(703) 898-7000

OWENS-CORNING ROOFING PRODUCTS
Fiberglas Tower
Toledo, OH 43659
(800) 342-3745

PABCO ROOFING PRODUCTS
1718 Thorne Rd.
Tacoma, WA 98421
(206) 272-0374

TARMAC ROOFING SYSTEMS INC.
E-Z Roll Roofing
1401 Silverside Rd.
Wilmington, DE 19810
(302) 475-7974

Caulks, Coatings, Sealants, Films, and Flashings

AKZO NOBEL COATINGS, INC.
(coil coatings for metal roofing)
P.O. Box 489
1313 Windsor Ave
Columbus, OH 43216
(614) 294-3361

ALBION INDUSTRIAL PRODUCTS
(caulks and sealants)
2195 Ekers Ave.
Montreal, H3S 1C7
Canada
(514) 737-2723

CABOT STAINS
100 Hole St.
Newburyport, MA 01940
(508) 465-1900 or (800) 877-8246

CHENEY FLASHING COMPANY
623 Prospect St.
Box 818
Trenton, NJ 08605-0818
(800) 322-2873

DAP PRODUCTS INCORPORATED
P.O. Box 277
Dayton, OH 44501
(513) 667-4461

EASY HEAT, INC.
(snow-melt cables)
31975 U.S. 20 E.
New Carlisle, IN 46552
(219) 654-3144

ELF ATOCHEM NORTH AMERICA, INC.
(Fluoropolymers: metal roof coatings)
2000 Market St.
Philadelphia, PA 19103
(215) 419-7520

ELK CORPORATION
(Prestique: roof accessory paint)
14643 Dallas Parkway
Suite 1000
Dallas, TX 75240
(214) 851-0400

FIBATAPE ROOF REPAIR REINFORCEMENT FABRIC
(PermaGlas-Mesh)
P.O. Box 220
Dover, OH 44622
(800) 762-6694

FIELDS CORPORATION
(roof tile cement)
2240 Taylor Way
Tacoma, WA 98421
(206) 627-4098

GIBSON-HOMANS COMPANY
1755 Enterprise Parkway
Twinsburg, OH 44087
(216) 425-3255

HARDCAST INC.
(Aluma-Grip-701: flexible flashing tape)
Box 1239
Wylie, TX 75098
(214) 442-6545

MACCO ADHESIVES
30400 Lakeland Blvd.
Wickliffe, OH 44092
(216) 943-6161

MINNESOTA MINING & MANUFACTURING
COMPANY
Construction Markets Division
3M Scotchtint Skylight & Window Films
3M Center, Building 225-4S-08
St. Paul, MN 55144-1000
(800) 362-3456

MOHECO PRODUCTS CO.
(metal flashings)
26835 W. Seven Mile St.
Detroit, MI 48240
(800) 959-0837

MONIER
(pre-formed plastic flashings)
P.O. Box 5567
Orange, CA 92613-5567
(714) 750-5366

MORTON INDUSTRIAL COATINGS
(metal roofing)
100 North Riverside Plaza
Chicago, IL 60606
(312) 807-3487

MULE-HIDE COATINGS
(metal protective roof coatings)
2924 Wyetta Dr.
Beloit, WI 53511
(800) 786-1492

NORTON PERFORMANCE PLASTICS CORP.
1 Sealants Park
Granville, NY 12832
(800) 724-0883

OLYMPIC STAINS
PPG Architectural Finishes, Inc.
1 PPG Place
Pittsburgh, PA 15272
(800) 441-9695

OREGON RESEARCH AND
DEVELOPMENT CORP.
(reflective roof sealants, tapes,
and coatings)
1895 16th St. S.E.
Salem, OR 97302
(800) 345-0809

P.L. ADHESIVES & SEALANTS
889 Valley Park Dr.
Shakopee, MN 55379
(800) 828-0253

PREMIUM PLASTICS, INC.
465 W. Cermak Rd.
Chicago, IL 60616
(312) 225-8700

PORTALS PLUS, INC.
(roof penetration products)
639 Thomas Dr.
Bensonville, IL 60106
(800) 624-8642

RED DEVIL INC.
2400 Vauxhall Rd.
Union, NJ 07083
(908) 688-6900

VENTURE TAPE CORP.
30 Commerce Rd.
Rockland, MA 02370
(617) 331-5900

YORK MANUFACTURING, INC.
(flashing)
P.O. Box 1009
Sanford, ME 04073
(207) 324-1300

ZALESKI SNOWGUARDS
AND ROOFING SPECIALTIES
11 Aiden St.
New Britain, CT 06053
(203) 225-1614

ZYNOLYTE PRODUCTS, CO.
(Standard Brands Paint, Co.)
2320 E. Dominquez St.
P.O. Box 6244
Carson, CA 90749
(310) 513-0700

Concrete, Clay, and Cement
Tiles and Shingles,
Ceramic Shingles and Slates,
and Polymer Tiles

AMERICAN CEMWOOD CORPORATION
(Cemwood Shake: Portland cement
and wood fiber composite)
P.O. Box C
Albany, OR 97321
(800) 367-3471

AMERICHEM INC.
8280 College Parkway, Suite 204
Fort Myers, FL 33919
(813) 481-4980

BARTILE ROOFS, INC.
725 No. 1000 W.
Centerville, UT 84014
(800) 933-5038

BENDER ROOF TILEIND, INC.
P.O. Box 190
Belleview, FL 34421
(800) 888-7074 or (904) 245-7074

CERTAINTEED CORPORATION
Celadon Ceramic Slate
P.O. Box 309
New Lexington, OH 43764
(800) 235-7528, (800) 699-9988,
or (800) 782-8777

DELEO CLAY TILE
600 Chaney St.
Lake Elsinore, CA 92530
(800) 654-1119 or (909) 674-1578

EAGLE ROOFING PRODUCTS
(lightweight concrete tile)
3546 N. Riverside Ave.
Rialto, CA 92377
(800) 300-3245 or (909) 355-7000

ETERNIT
Excelsior Industrial Park
P.O. Box 679
Blandon, PA 19510
(800) 235-3155

EVEREST ROOFING PRODUCTS
(lightweight polymer roofing tiles)
2500 Workman Mill Rd.
Whitier, CA 90601
(800) 767-0267

FIBRECEM CORPORATION
P.O. Box 411368
Charlotte, NC 28241
(800) 346-6147

G. E. PLASTICS
1 Plastics Ave.
Pittsfield, MA 01201
(800) 845-0600 or (413) 448-7110

GLADDING, MCBEAN
P.O. Box 97
Lincoln, CA 95648
(916) 645-3341

JAMES HARDIE BUILDING PRODUCTS
10901 Elm Ave.
Fontana, CA 92337
(800) 942-7343 or (800) 426-4051

or

JAMES HARDIE BUILDING PRODUCTS
#203-1182 Welch St.
North Vancouver, British Columbia V7P1B2
Canada

IMPRESSION
22599 S. Western Ave.
Torrance, CA 90501
(310) 618-1299

INTERNATIONAL ROOFING PRODUCTS, INC.
1832 South Brand Blvd., #200
Glendale, CA 91204
(818) 382-3514

LIFETILE CORPORATION
3511 N. Riverside Ave.
Rialto, CA 92376
(800) 562-8500 or (714) 822-4407

LUDOWICI-CELADON, INC.
P.O. Box 59
4757 Tile Plant Rd.
New Lexington, OH 43764
(800) 945-8453 or (614) 342-1995

MARLEY ROOF TILES
1990 E. Riverview Dr.
San Bernardino, CA 92408
(800) 344-2875 or (714) 796-8324

MARUHACHI CERAMICS OF AMERICA, INC.
1985 Sampson Ave.
Corona, CA 91719
(909) 736-9590

MAXITILE, INC.
17141 S. Kingsview Ave.
Carson, CA 90746
(800) 338-8453
or (310) 217-0316

MONIER ROOF TILE RE-ROOF DIVISION
1990 E. Riverview Dr.
San Bernardino, CA 92408
(800) 273-7663

MONIER, INC.
(villa and mission roofing tiles)
1832 S. 51 Ave.
Phoenix, AZ 85043
(800) 845-9921

or

MONIER, INC.
1745 Sampson Ave.
Corona, CA 91719
(800) 344-2875 or (714) 750-5366

NELCO ENGINEERING, INC.
(lightweight thermoplastic resin
roof panels)
4923 Route 100
New Tripoli, PA 18066
(610) 298-3125 or G.E. Plastics
(800) 845-0600

PIONEER CONCRETE TILE
(lightweight concrete shakes)
10650 Popular Ave.
Fontana, CA 92337
(909) 350-4238

POLETTO AND HAUCK
Contractors and Distributors, Inc.
432 West Gay St.
West Chester, PA 19380
(800) 537-4138

RE-CON BUILDING PRODUCTS, INC.
(fiber-cement shakes, slates,
and shingles)
P.O. Box 1094
Sumas, WA 98295
(800) 347-3373

or

RE-CON BUILDING PRODUCTS, INC.
33610 East Broadway Ave.
Mission, British Columbia V2V 4M4
Canada

SUPRADUR MANUFACTURING CORP.
P.O. Box 908
411 Theodore Fremd Ave.
Rye, NY 10580
(800) 223-1948 or (914) 967-8230

US TILE
909 West Railroad St.
Corona, CA 91720
(909) 737-0200

VANDE HEY'S ROOFING TILE CO., INC.
1665 Bohm Dr.
P.O. Box 263
Little Chute, WI 54140
(800) 537-4138

WESTILE
8311 W. Carder Court
Littleton, CO 80125
(800) 433-8453 or (303) 791-1711

Decking, Insulation,
and Sealing Materials

AMOCO FOAM PRODUCTS COMPANY
(insulation and housewrap)
2907 Log Cabin Dr.
Smyrna, GA 30080
(800) 241-4402

BRANCH RIVER FOAM PLASTICS, INC.
15 Thurbers Blvd.
Smithfield, RI 02917
(401) 232-0270

CELLOFOAM NORTH AMERICA, INC.
Perma-Vent Ventilation Channels
P.O. Box 406
Conyers, GA 30207
(800) 241-3634

CELOTEX FOAM SHEATHINGS
P.O. Box 31602
Tampa, FL 33631
(800) 235-6329

CORNELL CORPORATION
888 S. 3rd St.
P.O. Box 338
Cornell, WI 54732 (715) 239-6411

DOW CHEMICAL COMPANY
Construction Materials Group
(Styrofoam residential sheathing)
2020 Dow Center
Midland, MI 48674
(800) 232-2436 or (800) 441-4369

DUPONT
Tyvek Housewrap
P.O. Box 80705
Wilmington, DE 19880-0705

ENVIRONMENTALLY SAFE PRODUCTS, INC.
(roof deck insulation)
313 West Golden Ln.
New Oxford, PA 17350
(717) 624-3581

GAF CORPORATION
Weather Watch (Ice-&-water shield
membrane)
1361 Alps Rd.
Wayne, NJ 07470-3689
(201) 628-3000

HOMASOTE COMPANY
(residential roof sheathing and insulation)
P.O. Box 7240
West Trenton, NJ 08628
(800) 257-9491

KOOL*PLY RADIANT BARRIER DECKING
10616 Hempstead Rd.
Building F
Houston, TX 77092
(713) 680-8840

KOPPERS INDUSTRIES, INC.
Koppers Building
436 Seventh Ave.
Pittsburgh, PA 15219
(800) 558-2706

NRG BARRIERS
27 Peal St.
Portland, ME 04101
(800) 343-1285

RAVEN INDUSTRIES
(vapor barriers & housewrap)
P.O. Box 5107
Sioux Falls, SD 57117
(800) 635-3456

REEF INDUSTRIES, INC.
Griffolyn (roof deck vapor barriers)
P.O. Box 750250
Houston, TX 77275
(800) 231-6074

TYPAR HOUSEWRAP
70 Old Hickory Blvd.
P.O. Box 511
Old Hickory, TN 37138
(800) 321-6271

UCI INDUSTRIES INC.
3 Century Dr.
Parsippany, NJ 07054
(800) 828-7155

VELUX-AMERICA, INC.
(skylight roofing underlayment)
P.O. Box 5001
450 Old Brickyard Rd.
Greenwood, SC 29648-5001
(800) 888-3589

Fasteners, Nails, Power Nailers and Staplers, and Staples

AIRY SALES CORP.
(staplers)
14535 Valley View Ave., #N
Santa Fe Springs, CA 90670
(310) 926-6192

APACHE HOSE & BELTING, INC.
(gas- and electric-powered compressors)
2515 18th St. S.W.
P.O. Box 1719
Cedar Rapids, IA 52406
(319) 365-0471

ARROW FASTENER COMPANY, INC.
271 Mayhill St.
Saddle Brook, NJ 07662
(201) 843-6900

BOSCH POWER TOOLS
S-B Power Tool Company
4300 West Peterson Ave.
Chicago, IL 60646
(312) 286-7330

CHICAGO PNEUMATIC TOOL CO.
2200 Bleecker St.
Utica, NY 13501
(315) 792-2600

DUO-FAST CORPORATION
702 N. River Rd.
Franklin Park, IL 60131
(708) 678-0100

J & M FASTENERS
181-B Mayhew Way
Walnut Creek, CA 94596
(800) 727-8698 or (510) 932-4484

FASTENING SYSTEMS WAREHOUSE
3486 Walden Ave.
Depew, NY 14043
(800) 944-1556

GLENDENIN BROS., INC.
(copper nails)
4309 Erdman Ave.
Baltimore, MD 21213

GRIZZLY IMPORTS, INC.
Tool Catalog Sales
1821 Valencia St.
Bellingham, WA 98226
(800) 523-4777 or (206) 647-0801

or

GRIZZLY IMPORTS, INC.
Tool Catalog Sales
2406 Reach Rd.
Williamsport, PA 17701
(800) 541-5537 or (717) 326-3806

HITACHI POWER TOOLS
3950 Steve Reynolds Blvd.
Norcross, GA 30093
(800) 706-7337 or (404) 925-1774

KEYSTONE STEEL & WIRE COMPANY
Red Brand Nails
7000 South West Adams
Peoria, IL 61641
(309) 697-7020

KOPPERS INDUSTRIES, INC.
Koppers Building
436 Seventh Ave.
Pittsburgh, PA 15219
(800) 558-2706

MANASQUAN
(stainless steel fasteners)
P.O. Box 669
Allenwood, NJ 08720-0669
(800) 542-1978

MAZE NAILS
Division of W.H. Maze Company
P.O. Box 449
100 Church St.
Peru, IL 61354
(800) 435-5949 or (815) 223-8290

MAKITA U.S.A., INC.
14390 Northam St.
La Mirada, CA 90638
(800) 462-5482

MODI-SYSTEMS
(propane torch)
275 Forest Ave., Suite 206-A
Paramus, NJ 07652
(800) 222-6634 or (201) 599-0604

NEWPORT FASTENER
1300 Gene Autry Way
Anaheim, CA 92805
(714) 385-1111

OLDHAM UNITED STATES SAW CORP.
(roofers' carbide blades)
P.O. Box 10
Burt, NY 14028
(716) 778-8588

PASLODE
(cordless fasteners)
Lincolnshire, IL 60009
(800) 323-1303

PHIFER WIRE PRODUCTS, INC.
(aluminum nails)
P.O. Box 1700
Tuscaloosa, AL 35403
(205) 345-2120

PNEUMATIC SPECIALTY PRODUCTS
37250 PLYMOUTH RD.
Livonia, MI 48150
(313) 591-6309

PRUDENTIAL METAL SUPPLY
CORPORATION
(copper nails)
171 Milton St.
East Dedham, MA 02026
(617) 329-2800

SENCO PRODUCTS, INC.
(pneumatic staplers)
8485 Broadwell Rd.
Cincinnati, OH 45244
(800) 543-4596 or (513) 388-2000

SPOTNAILS
Division of Peach Industries, Ltd.
1100 Hicks Rd.
Rolling Meadows, IL 60008
(800) 873-2239 or (708) 259-1620

STANLEY BOSTITCH, INC.
Fastening Systems
Briggs Dr.
East Greenwich, RI 02818
(401) 884-2500

SWAN SECURE PRODUCTS, INC.
(copper and stainless steel
roofing nails)
1701 Parkman Ave.
Baltimore, MD 21230
(410) 646-2800

SWINGLINE FASTENERS
3200 Skillman Ave.
Long Island City, NY 11101
(718) 729-9600

TECO WOOD FASTENERS
Colliers Way
Colliers, WV 26035
(800) 438-8326

TREMONT NAIL COMPANY
(old-fashioned cut nails)
21 Elm St.
Box 111
Wareham, MA 02571
(508) 295-0038

VACO TOOLS & FASTENERS
1510 Skoki Blvd.
Northbrook, IL 60062

THE WIRE WORKS, INC.
(slate and tile roof fasteners
and hardware)
P.O. Box 639
Virginia City, NV 89440
(800) 341-8828

Hand Tools

AJC HATCHET CO.
1227 NORTON RD.
Hudson, OH 44236
(800) 428-2436 or (216) 655-2851

AARON INDUSTRIAL SUPPLY
453 W. 28th St.
Hialeah, FL 33010
(305) 888-8877

AMERICAN TOOL COMPANIES, INC.
Prosnip Aviation Snips
P.O. Box 337
DeWitt, NB 68341
(402) 683-2315

AMES TOOLS
Box 1774
Parkersburg, WV 26102
(800) 624-2654

ARDELL INDUSTRIES, INC.
P.O. Box 1573
555 Lehigh Ave.
Union, NJ 07083
(908) 687-5900

COOPER INDUSTRIES
Hand Tool Division
3012 Mason St.
Monroe, NC 28110
(704) 289-8486

ESTWING MFG. CO.
2647 8th St.
Rockford, IL 61109
(815) 397-9558

EVERGREEN SLATE CO., INC.
(slating tools)
P.O. Box 248
68 East Potter Ave.
Granville, NY 12832
(518) 642-2530

HART TOOL COMPANY
(hammers and measuring tapes)
P.O. Box 862
Fullerton, CA 92632
(800) 331-4495

HYDE MFG. CO.
Dept. D.
Eastford Rd.
Southbridge, MA 01550
(800) 872-4933

KESON INDUSTRIES, INC.
1660 Quincy Ave.
P.O. Box 394
Naperville, IL 60566
(708) 369-8848

LUMARK INDUSTRIAL PRODUCTS CORPORATION
(tool belts)
P.O. Box 6262
Parsippany, NJ 07054
(800) 544-5297

McGUIRE-NICHOLAS MFG. CO.
2331 S. Tubeway Ave.
Commerce, CA 90040
(213) 722-6961

MIDWEST OHIO TOOL COMPANY
P.O. Drawer 269
Upper Sandusky, OH 43351
(419) 294-1987

MOHECO PRODUCTS CO.
(magnetic rake)
26835 W. Seven Mile St.
Detroit, MI 48240
(800) 959-0837

NAILERS, INC.
(tool belts, pouches, etc.)
10845-C Wheatlands Ave.
Santee, CA 92071
(619) 562-2215

STANLEY WORKS
P.O. Box 7000
600 Myrtle St.
New Britain, CT 06050
(800) 648-7654 or (203) 225-5111

L.S. STARRETT CO.
121 Crescent St
Athol, MA 01331
(508) 249-3551

U.S. GENERAL
Tool Catalog
100 Commercial St.
Plainview, NY 11803
(516) 576-9100

VAUGHAN & BUSHNELL MFG. CO.
P.O. Box 390
11414 Maple Ave.
Hebron, IL 60034
(815) 648-2446

WHOLE EARTH ACCESS
822 Anthony St.
Berkeley, CA 94710
(800) 829-6300

Ladders, Scaffolds, and Powered Hoists

ADVANCED DESIGN PRODUCTS
P.O. Box 447
Finksburg, MD 21048
(800) 743-8815 or (410) 833-8814

ALACO LADDER COMPANY
5167 G. St.
Chino, CA 91710
(714) 591-7561

ALUM-A-POLE CORP.
(pump jacks and scaffolding)
2589 Richmond Terrance
P.O. Box 30066
Staten Island, NY 10303
(718) 447-2608

BAUR CORPORATION
1505 East Bowman St.
Wooster, OH 44691
(216) 262-3070

HOWARD MANUFACTURING COMPANY
P.O. Box 1188
Kent, WA 98035
(206) 852-0640

KELLER LADDERS
18000 State Rd. Nine
Miami, FL 33162
(305) 651-7100

or

KELLER LADDERS
Magic Mall Industrial Park
Swansboro, GA 30401
(912) 237-7504

THE LADDER WORKS, INC.
270 N. Eisenhower Lane
Lombard, IL 60148
(800) 572-5462

LOUISVILLE LADDER DIVISION
Emerson Electric Company
1163 Algonquin Parkway
Louisville, KY 40208
(800) 666-2811

REIMANN & GEORGER, INC.
(hoisting and material handling equipment)
P.O. Box 681
1849 Harlem Rd.
Buffalo, NY 14240
(716) 895-1156

R.D. WERNER CO., INC.
93 Werner Rd.
Greenville, PA 16125
(800) 221-7307

WING ENTERPRISES, INC.
P.O. Box 3100
Springville, UT 84663
(800) 453-1192

SAFETY LIFT COMPANY
(material handling equipment)
P.O. Drawer 325
Columbiana, OH 44408
(216) 482-3861

SMITH HOIST MFG., CO., INC.
(portable hoists)
P.O. Box 205
Cassopolis, MI 49031
(616) 445-2433

Metal Roofing

AEICOR METAL ROOFING PRODUCTS
COMPANY, INC.
450 W. McNab Rd.
Ft. Lauderdale, FL
(305) 974-3300

AEP-SPAN
5100 E. Grand Ave.
P.O. Box 150449
Dallas, TX 75315
(800) 527-2503

ALCAN BUILDING PRODUCTS
Fabral Metal Roofing
P.O. Box 4608
3449 Hempland Rd.
Lancaster, PA 17604
(717) 397-2741

ALCOA BUILDING PRODUCTS, INC.
P.O. Box 716
Sidney, OH 45365
(800) 621-7466 or (800) 962-6973

ARS, INC.
#8 50th St. North
Birmingham, AL 35212
(205) 591-5105

ASC PACIFIC, INC.
2110 Enterprise Blvd.
West Sacramento, CA 95691
(800) 726-2727

ATLANTA METAL PRODUCTS, INC.
5700 Riverview Dr.
Mableton, GA 30059
(800) 554-1097

ATLAS ALUMINUM CORP.
6612 Snowdrift Rd.
Allentown, PA 18106
(800) 468-1441 or (610) 395-8445

BERRIDGE MANUFACTURING COMPANY
1720 Maury St.
Houston, TX 77026
(800) 231-8127

BINKLEY BUILDING PRODUCTS DIVISION
11970 Borman Dr., Suite 200
St. Louis, MO 63146
(314) 434-7110

CARTER HOLT HARVEY ROOFING USA, INC.
(lightweight steel-panel roofing)
2600 South Loop West, Suite 230
Houston, TX 77054
(713) 913-4032

CECO BUILDINGS INC.
(retrofit metal roofing)
P.O. Box 6500
Highway 45 North
Columbus, MS 39703
(601) 328-6722

CINDU METAL ROOFING PANELS
12591 Perimeter Dr.
Dallas, TX 75228
(800) 879-1172

COLORKLAD VINCENT METALS
P.O. Box 360
Minneapolis, MN 55440
(800) 328-7800

COPPER SALES, INC.
Una-Clad Roofing
1405 North Highway 169
Minneapolis, MN 55441
(800) 426-7737

DELEOA INDUSTRIES, INC.
1950 Northwest 18th St.
Pompano Beach, FL 33069
(305) 979-5004

ECI BUILDING COMPONENTS, INC.
P.O. Box 968
Stafford, TX 77497
(800) 669-9324

ENGLERT INCORPORATED
1200 Amboy Ae.
Peth Amboy, NJ 08862
(908) 826-8614

FOLLANSBEE STEEL CORPORATION
(terne roofing products)
Box 610
Follansbee, WV 26037
(800) 624-6906 or (304) 527-1260

FOREMOST MANUFACTURING CO.
P.O. Box 52549
Livonia, Mi 48152
(800) 622-4001

GERARD ROOFING TECHNOLOGIES
(lightweight, stone-coated steel tiles
and shakes)
955 Columbia St.
Brea, CA 92621
(800) 237-6637, (800) 841-3213, or (714)
529-0407

W.P. HICKMAN COMPANY
Microzinc Roofing System
P.O. Box 15005
Asheville, NC 28813
(800) 438-3897 or (704) 274-4000

HOMETECH STEEL
(steel-frame homes)
10579 Dale St., Suite 110
Stanton, CA 90680-1175
(800) 600-0009

JAMES RIVER STEEL, INC.
P.O. Box 11498
Richmond, VA 23230
(800) 825-0717

MBCI
(metal roof and wall systems)
14031 West Hardy
P.O. Box 38217
Houston, TX 77238
(713) 445-8555

MCELROY METAL, INC.
(residential roofing products)
P.O. Box 1148
Shreveport, LA 71163
(800) 950-6531

MERCHANT & EVANS, INC.
P.O. Box 1680
100 Connecticut Dr.
Burlington, NJ 08016
(800) 257-6215

METAL SALES MANUFACTURING CORPORATION
(polyester-coated, galvanized steel roof
panels)
7800 State Rd. 60
Sellersburg, IN 47172
(800) 944-3786

MET-TILE, INC.
(metal tile panel roofing)
P.O. Box 4268
Ontario, CA 91761
(800) 899-0311 or (909) 947-0311

REINKE SHAKES
3321 Willowwood Circle
Lincoln, NE 68506
(800) 228-4312

REVERE COPPER PRODUCTS, INC.
(cooper roofing)
P.O. Box 300
Rome, NY 13442
(800) 448-1776 or (315) 338-2022

REYNOLDS METALS COMPANY
Construction Products
P.O. Box 27003
Richmond, VA 23261
(804) 281-4188

STEEL TILE CO.
Metal Roofing Systems
R.R.1 Thornton, Ontario LOL 2NO
Canada
(705) 436-1723

TEGOLA USA
3807 Inwood Landing
Orlando, FL 32812
(800) 545-4140

THERMO MATERIALS, INC.
(copper roofing)
P.O. Box 9454
San Diego, CA 92109
(800) 882-7007

UNITED STATES STEEL CORPORATION
600 Grant St.
Pittsburgh, PA 15230
(412) 433-6864

VICWEST STEEL
9000 Wessex Pike
Suite 201
Louisville, KY 40222
(502) 339-7222

VINCENT METALS
Building Products Division
724 24th Ave., SE
Minneapolis, MN 55414
(800) 328-7772 or (612) 378-1131

WALCON CORP.
4375 Second St.
Ecorse, MI 40229
(313) 382-4000

WEIRTON STEEL CORP.
400 Three Springs Dr.
Weirton, WV 26062
(800) 223-9777

WHEELING-PITTSBURGH STEEL
CORPORATION
Wheeling Corrugating Co.
1134-40 Market St.
Wheeling, WV 26003
(304) 234-2773

ZAPPONE MANUFACTURING
N. 2928 Pittsburg
Spokane, WA 99207
(509) 483-6408

Photovoltaics

KANEKA CORP.
3-2-4 Nakanoshima Kita-Ku
Osaka 530
Japan
06-226-5050

KUBOTA CORP.
1-2-47 Shikitsu-Higashi Naniwa-Ku
Osaka 556
Japan
06-648-2111

KYOCERA CORP.
5-22 Kita-Inouecho
Higashino, Yamashina-Ku
Kyoto 607
Japan
075-592-3851

MISAWA HOMES CO. LTD.
2-4-5, Takaido-Higashi, Suginami-Ku
Tokyo 168
Japan
Fax 03-3349-8074

SANYO ENERGY CORP.
Amorton Solar Cells
2001 Sanyo Ave.
San Diego, CA 92173
(619) 661-6020

U.S. DEPARTMENT OF ENERGY
Conservation and Renewable Energy
Photovoltaics Technology Division
Room 5H-088
1000 Independence Ave., SW
Washington, D.C. 20585
(202) 586-1720

Publications and Videocassetes

ABC SUPPLY COMPANY INC.
Attn: Catalog Division
P.O. Box 838
Beloit, WI 53512
(800) 366-2227

BUILDER MAGAZINE
1996 Guide to Building Products
(CD-ROM; $39.95)
Hanely-Wood Inc.
8420 W. Gryn Mawr Ave.
Chicago, IL 60631
(800) 241-2537 or (312) 380-9000

CONSUMER INFORMATION CATALOG
Pueblo, CO 81009

GEORGIA-PACIFIC CORPORATION
(videocassette: "Your Guide to Reroofing
Your Home")
133 Peachtree St., NW
Atlanta, GA 30303
(800) 284-5347

"HOMETIME"
videocassette: "Roofing Preparation
& Installation"; price: $9.95)
4275 Norex Dr.
Chaska, MN 55318
(800) 992-4888 or (800) 382-1991

MID-AMERICA BUILDING PRODUCTS CORP.
videocassette: "Soffit, Ridge &
Hip Ventilators"; $9.95)
9246 Hubbell Ave.
Detroit, MI 48228
(800) 521-8486

NATIONAL ASSOCIATION
OF HOME BUILDERS
Home Builder Press
15th and M St.'s, NW
Washington, D.C. 20005
(800) 368-5242

OWENS-CORNING ROOFING PRODUCTS
(videocassette: "How to Install
Shingles: Nailing Down the Basics")
Fiberglas Tower
Toledo, OH 43659
(800) 342-3745

ROOFER—EQUIPMENT DIRECTORY ISSUE
(Price: $6.00)
D&H Publications, Inc.
6719 Winkler Rd., #214
Fort Myers, FL 33919
(813) 459-2929

ROOFER MAGAZINE
D&H PUBLISHING
10990 Metro Parkway St.
Ft. Myers, FL 33912
(813) 275-7663

ROOFING MATERIALS GUIDE
(Price: $95.00 per year)
National Roofing Contractors Association
Box 4752
North Suburban, IL 60197
(708) 299-9070

ROOFING/SIDING/INSULATION—TRADE
DIRECTORY ISSUE
(Price: $25.00, includes shipping)
Advanstar Communications, Inc.
7500 Old Oak Blvd.
Cleveland, OH 44130
(216) 243-8100

SUPERINTENDENT OF DOCUMENTS
U.S. Government Printing Office
Washington, DC 20402
(202) 783-3238

TRIMLINE
(free informational videocassette:
"Shingle-Over-Ridge Vent")
705 Pennsylvania Ave.
Minneapolis, MN 55426
(800) 438-2920

UNIFORM BUILDING CODE
International Conference of Building
Officials
5360 Workman Mill Rd.
Whittier, CA 90801
(310) 699-0541

U.S. DEPARTMENT OF COMMERCE
National Technical Information Service
Springfield, VA 22161
(703) 487-4650)

VERMONT STRUCTURAL SLATE COMPANY, INC.
Slate Roofs (price: $11.95)
3 Prospect St.
Box 98
Fair Haven, VT 05743
(800) 343-1900 or (802) 265-4933

Recyclers

GAF CORPORATION
Pallets Plus (wood-pallet recycling
program)
Wayne, NJ 07470-3689
(201) 628-3000

RECLAIM INC.
(recycles worn asphalt shingles into
street-paving and road-patching materials)
8001 N. Dale Maybry
Tampa Bay, FL 33614
(813) 935-8533

Safety Equipment

Cabot Safety Corp.
(safety eyewear)
90 Mechanics St.
Southbridge, MA 01550
(508) 764-5500

Miller Equipment Division
WGM Saftey Corp.
(fall protection)
1355 15th St.
P.O. Box 271
Franklin, PA 16323
(814) 432-2118

Rose Manufacturing Co.
(fall protection)
2250 South Tejon St.
Englewood, CO 80110
(800) 722-1231 (303) 922-6246

RTC
(fall protection)
3101 Market St
Wilmington, DE 19802
(302) 762-4300

The Sinco Group
(fall protection)
One Sinco Pl.
P.O. Box 361
East Hampton, CT 08424
(800) 864-2699

Skylights and Roof Windows

Acralight Skylights
2491 Du Bridge Ave
Irvine, CA 92714
(800) 273-9343

Anderson Windows, Inc.
P.O. Box 3900
Peoria, IL 61614
(800) 426-4261

APC Corporation
50 Utter Ave.
Hawthorne, NJ 07506
(800) 222-0201

Bristolite Skylights
401 E. Goetz Ave.
P.O. Box 2515
Santa Ana, CA 92707
(800) 854-8618

Eagle Window & Door
375 E. 9th St.
P.O. Box 1072
Dubuque, IA 52004-1072
(319) 556-3825

Flore Skylights, Inc.
700 Grace St.
Somerdale, NJ 08083
(800) 346-7310

Gordon Skylights
133 Mata Way, Suite 101
San Marcos, CA 92069
(619) 727-2008

Hillsdale Industries
5049 South National Dr.
Knoxville, TN 37914
(615) 637-1711

MARVIN WINDOWS AND DOORS
Highway 11 W
WarRd., MN 56763
(800) 346-5128

MID-AMERICA BUILDING PRODUCTS CORP.
(soffit, ridge & hip ventilators)
45657 Port St.
Plymouth, MI 48170
(800) 521-8486

NATURALITE, INC.
P.O. Box 629
750 Airport Rd.
Terrell, TX 75160
(800) 527-4018

NATURLITE SUN TUBES
4745 126 Avenue North, Suite 42
Clearwater, FL 34622
(800) 839-2900 or (813) 573-1474

BENJAMIN OBDYKE, INC.
John Fitch Industrial Park
Warminister, PA 18974
(800) 523-5261

ODL INCORPORATED
215 E. Roosevelt Ave.
Zeeland, MI 49464
(800) 288-1800

ONDULINE ROOFING PRODUCTS
4900 Onduline Dr.
Fredericksburg, VA 22401
(703) 898-7000

PELLA SKYLIGHTS
102 Main St.
Pella, IA 50219
(800) 845-4525

PLASTECO, INC.
P.O. Box 24158
Houston, TX 77229
(713) 453-8696

ROLLAMATIC ROOFS INC.
1441 Yosemite Ave.
San Francisco, CA 94124
(415) 822-5655

SKYLINE SKY-LITES
2903 Delta Dr.
Colorado Springs, CO 80910
(800) 759-9046

SKYMASTER SKYLIGHTS
413 Virginia Dr.
Orlando, FL 32803

SOLATUBE
5825 Avenida Encinas, Suite 101
Carlsbad, CA 92008
(800) 773-7652

SUN TUNNEL
FreeLite, Inc.
Reseda, CA 91335
(800) 369-3664
or (818) 727-0050

or

SUN TUNNEL
FreeLite, Inc.
11837 Judd Ct.
Dallas, TX 75243
(214) 234-5990

SUNGLO SKYLIGHT PRODUCTS
3124 Gillham Plaza
Kansas City, MO 64109
(800) 821-6656

THERMO-VU SKYLIGHTS
51 Rodeo Dr.
Edgewood, NY 11717
(516) 243-1000
or (800) 883-5483

VELUX-AMERICA, INC.
P.O. Box 5001
450 Old Brickyard Rd.
Greenwood, SC 29648-5001
(800) 888-3589

Ventarama Skylights
303 Sunnyside Blvd.
Plainview, NY 11803
(800) 237-8096 or (516) 931-0202

Wasco Products, Inc.
Residential Division
P.O. Box 351
Pioneer Ave.
Sanford, ME 04073
(800) 866-8101 or (800) 338-1181

Slate Roofing

Sam F. Berger Co.
P.O. Box 6053
Ashland, VA 23005

E.R. Blaisdell Slate Products Company
27 Osgood St., Box 29
Somerville, MA 02143
(617) 666-5720

H.C. Brandt Distributing
P.O. Box 1426
848 S. Myrtle Ave., Unit #5
Monrovia, CA 91016
(818) 303-3919

Echeguren Slate, Inc.
1465 Illinois St.
San Francisco, CA 94107
(800) 992-0701

Hilltop Slate, Inc.
Route 22-A
Middle Granville, NY 12849
(518) 642-2270

Buckingham Virginia Slate Corporation
Box 8
1 Main St.
Arvonia, VA 23004
(804) 581-1131

Building & Industrial Wholesale
Company
P.O. Box 1038
Parkersburg, WV 26101

Central States Slate Sales
3105 West North Ave.
Milwaukee, WI 53208
(414) 873-7885

Robert Cunningham & Company
699 Second St.
San Francisco, CA 94107
(415) 542-7777

Domestic Marble & Stone Corp.
41 East 42nd St.
New York, NY 10017

Durable Slate Co.
1050 N. Fourth St.
Columbus, OH 43201
(800) 666-7445

Emack Slate Company, Inc.
9 Office Park Circle
Suite 120
Birmingham, AL 35223
(205) 879-3424

Evergreen Slate Co., Inc.
P.O. Box 248
68 East Potter Ave.
Granville, NY 12832
(518) 642-2530

Hamre Associates
P.O. Box 849
431 Wonder Dr.
White Rock, SC 29177
(803) 781-5834

HILLTOP SLATE, INC.
Route 22-A
Middle Granville, NY 12849
(518) 642-2270

INLAND STONE CORPORATION
480-1 Madison Ave.
Calumet City, IL 60409
(708) 891-4093

J.L.J. ENTERPRISES, LTD.
1008 Concordia Dr.
Baltimore, MD 21204

J.G.A. ATLANTA
2200 Cook Dr.
Doraville, GA 30340
(404) 447-6466

C.W. KERNPKAU, INC.
600 42nd Ave.
Nashville, TN 37209

MARYLAND STONE SERVICE, INC.
2418 Bradford Rd.
Baltimore, MD 21234

MONIER
Normandy Slate
P.O. Box 5567
Orange, CA 92513-5567
(714) 750-5366

NATURAL SLATE
40 N. School St.
Honolulu, HI 96817
(808) 533-2220

NEIL, INC.
1626 Harbeson Ave.
Cincinnati, OH 45224
(513) 541-3554

NEW ENGLAND SLATE CO.
Burr Pond Rd.
Sudbury, VT 05733
(802) 247-8809

NEW ENGLAND SLATE LIMITED
Effingham Rd. & Metier Rd.
Pelham, Ontario
Canada

NEWFOUNDLAND SLATE, INC.
8800 Sheppard Ave. E
Scarborough, Ontario M1B5R4
Canada

NORTH BANGOR SLATE COMPANY
N. 1st St.
Bangor, PA 18013
(215) 588-2154

OTTEY & HOOPES
P.O. Box G
Paoli, PA 19301

RISING & NELSON SLATE COMPANY
P.O. Box 98
Main St.
West Pawlet, VT 05775
(800) 348-8820
or (802) 645-0150

THE ROOF CENTER
5244 River Rd.
Bethesda, MD 20816
(301) 656-9231

ROOF TILE, ETC.
14611 Cyprus Point
Farmers Branch, TX 75234

ROOFERS MART OF VIRGINIA
3090 Aspen Ave.
Richmond, VA 23228
(804) 264-4747

SLATE INTERNATIONAL, INC.
15011 Marlboro Pike
Upper Marlboro, MD 20772
(800) 343-9785

SNOW LARSON, INC.
1925 Oakcrest Ave.
Roseville, MN 55405
(612) 636-0630

STRUCTURAL SLATE COMPANY
P.O. Box 187
222 E. Main St.
Pen Argyl, PA 18072
(215) 863-4141

VERMONT SLATEWRIGHT
RR 1, Box 2617
Vail Rd.
Bennington, VT 05201
(800) 442-3181

VERMONT STRUCTURAL SLATE COMPANY, INC.
Box 98
3 Prospect St.
Fair Haven, VT 05743
(800) 343-1900 or (802) 265-4933

JOE S. WILLIAMS COMPANY
P.O. Box 7892
3912 Hawthorne Rd.
Rocky Mount, NC 27801

Ventilators

ADO PRODUCTS
(ventilation baffles)
7357 Washington Ave., South
Edina, MN 55439
(800) 666-8191

AIR VENT INC.
A CertainTeed Company
4801 N. Prospect Rd.
Peoria Heights, IL 61614
(800) 247-8368

ALCOA BUILDING PRODUCTS
Roof Ventilation Division
P.O. Box 716
2615 Campbell Rd.
Sidney, OH 45365
(513) 498-6140 or (800) 967-6973

AMPCOR ROOF VENTS
111 Fellowship Rd.
Taylorsville, MS 39168
(601) 785-4711

ARVIN INDUSTRIES, INC.
Arvin Wind Turbines
P.O. Box 3000
Columbus, IN 47202
(812) 379-3000

AUBREY MANUFACTURING, INC.
6709 South Main St.
Union, IL 60180
(815) 923-2101

COR-A-VENT
16250 Petro Dr.
Mishawak, IN 46544
(800) 837-8368

HICKS STARTER VENT
(combined drip edge and vent)
124 Main St.
Westford, MA 01886
(508) 692-8811

LEIGH PRODUCTS, INC.
Ridge Vent and Attic Ventilators
Coopersville, MI 49404
(616) 837-8141

LOMANCO
P.O. Box 519
Jacksonville, AK 72076
(501) 982-6511

MIDGET LOUVER COMPANY
800 Main Ave.
Route 7
Norwalk, CT 06852

MILGARD MANUFACTURING, INC.
3800 - 136th St. N.E.
Marysville, WA 98271
(800) 562-0402 or (206) 659-0836

NAUTILUS
Roof-Mount Powered Attic Ventilators
Hartford, WI 53027
(414) 673-4345

NORTH AMERICAN BUILDING PRODUCTS, INC.
1201 E. Whitcomb
Madison Heights, NJ 48071
(800) 521-9920

NRG BARRIERS
ISO-Vent
27 Perry St.
Portland, ME 04101
(800) 343-1285

BENJAMIN OBDYKE, INC.
(roll vents for cedar and asphalt-shingle
ridge ventilation)
John Fitch Industrial Park
Warminister, PA 18974
(800) 523-5261

ROBBINS & MYERS, INC.
Hunter Ventilating and Circulating Fans
2500 Frisco Ave.
Memphis, TN 38114
(901) 743-1368

TRIMLINE ROOF VENTILATION SYSTEMS INC.
705 Pennsylvania Ave.
Minneapolis, MN 55426
(800) 438-2920 or (612) 540-9737

Wood Shingles and Shakes

CEDAR VALLEY SHINGLE SYSTEMS
943 San Felipe Rd.
Hollister, CA 95023
(408) 636-8110

C&H ROOFING, INC.
Box 2105
Lake City, FL 32056
(800) 327-8115

GREEN RIVER
Box 515
Sumas, WA 98295
(800) 663-8707

HISTORIC OAK ROOFING
Summit Towers, Suite 913
201 Locus St
Knoxville, TN 37902
(800) 321-3781

HOOVER TREATED WOOD PRODUCTS
P.O. Box 746
Thomson, GA 30824
(800) 832-9663

KOPPERS COMPANY
437 Seventh Ave.
Pittsburgh, PA 15219
(800) 468-9629

LIBERTY CEDAR
535 Liberty Lane
West Kingston, RI 02892
(800) 882-3327
or (401) 789-6626

LOUISIANA-PACIFIC CORP.
111 S.W. Fifth Ave.
Portland, OR 97204
(800) 547-6331
or (503) 221-0800

MASONITE DIVISION
Building Products Group
Woodruf Sales Department
1 South Wacker Dr.
Chicago, IL 60606
(800) 255-0785

BENJAMIN OBDYKE, INC.
(roll vent for cedar ridge ventilation and
Cedar Breather underlayment for cedar
roofs installed over plywood decks)
John Fitch Industrial Park
Warminister, PA 18974
(800) 523-5261

SOUTHCOAST SHINGLE COMPANY
2220 E. South St.
Long Beach, CA 90805
(310) 634-7100

SOUTH COUNTY POST AND BEAM
P.O. Box 432
W. Kingston, RI 02892
(401) 783-4415

SHAKERTOWN ROOFING AND SIDING
Box 400
1200 Kerron St.
Winlock, WA 98596
(800) 426-8970 or (206) 785-3501

TEAL CEDAR PRODUCTS
17835 Trigg Rd.
Surrey, British Columbia V3T 5J4
Canada

Glossary

abut To position snugly against the top or side of a shingle. Shingle courses can be abutted when a new layer is applied over worn but not warped or buckled shingles; butt.

alignment notch Factory-cut end of a shingle where cutouts meet to form proper shingle alignment.

angle roof brackets Metal supports—either fixed or adjustable to three positions—designed to provide a platform for planking on steeply pitched roofs.

anti-ponding metal Flashing installed at the eaves to provide a base for the initial course of tile roofing.

architectural shingles Asphalt-fiberglass shingles with a dense, multiple-layer, laminated construction to provide a shadow line designed to look like slate or a random edge resembling wood shingles. Heavyweight shingles usually sold in four or five bundles per 100 square feet of roof coverage and having a 35-year or 40-year warranty.

asphalt fiberglass-based shingles Roof shingles made with an inorganic fiberglass base saturated with asphalt and topped by colored ceramic granules or opaque rock for weather and sunlight resistance.

asphalt shingles Roof shingles made with an organic base, such as paper or wood chips, and felt saturated with asphalt and coated with colored ceramic granules or opaque rock for sunlight and weather resistance. Asphalt shingles have been supplanted in the market by asphalt fiberglass-based shingles.

attic ventilator Screened openings located in the soffit, gable ends, or at the ridge line. Power-driven fans can be used as part of a balanced exhaust system for homes.

aviation snips   *See* tin snips.

backing down   High-nailing courses of roofing material to tie in lower, successive course.

backing in   *See* filling in.

battens   Small strips of wood used as a foundation on which to hang tile, as support for lathing on which to hang tile, or to temporarily reinforce felt prior to installing shingles.

birdstops   On mission-barrel tile roofs, barriers placed at the eaves to prevent wind-driven moisture, birds, and insects from entering.

blind nailing   Installing nails so that the nail heads are concealed by roofing material.

bonded contractor   A roofing contractor who has appropriate business and liability insurance coverage.

border shingles   Roofing material applied to the outer edges of a roof section—such as rakes, eaves, and valleys—for protection and to present an even edge.

building paper   *See* felt.

build-up roofing   Applied to flat roofs with three to five layers of asphalt-saturated felt and roll roofing. Designed to hold water until it evaporates rather than shed water, hot tar is mopped and finished with crushed slag or gravel. Not appropriate for do-it-yourselfer installation; special equipment is required.

butt   *See* abut.

capillary action   On a low-slope roof surface, the attraction or repulsion of water due to the inherent surface tension of the solid roof surface and the liquid. The resulting backflow of rainwater does not easily run off the last few courses of shingles and into the gutters and has the potential to cause leaks.

capping shingles   One-tab shingles centered and applied horizontally to the ridge or hip of a roof; the last course of roll roofing centered and applied horizontally to the ridge.

caulk   Waterproofing material applied to chimney flashing seams or vent-pipe seams. Not a replacement for roofing cement.

cement tile roofing   *See* tile roofing.

ceramic granules   Finely crushed or ground minerals, sand, rock, or fire-hardened materials on asphalt-fiberglass shingles and roll roofing for color and weather resistance.

change order   After a contract is signed, a request by the customer or the contractor to, for example, substitute one brand, style, weight, or color of material for another, or to approve an adjustment of the estimated costs for labor or materials.

class A fire rating The highest fire-resistance rating assigned by Underwriters Laboratories for roofing materials, including clay, cement, slate, copper, and some composite materials.

class B fire rating The least frequently found rating for roofing materials assigned by Underwriters Laboratories.

class C fire rating The most common rating for roofing materials assigned by Underwriters Laboratories, including organic, composition shingles.

clay tile roofing *See* tile roofing.

closed valley Where two internally sloping sections intersect, shingles are woven instead of cut to form a channel.

coil nailer Pneumatic-powered roofing tool designed to deliver about 120 nails per load.

collar *See* flange.

composite roofing Roof material formed primarily from a combination of Portland cement and wood fiber to provide long-lasting, lightweight slates, shakes, and shingles.

composition shingles *See* asphalt shingles.

concrete tile roofing *See* tile roofing.

corrugated asphalt roofing Asphalt-impregnated tiles or sheets designed to be installed much like metal roofing panels.

counterflashing Metal strips used to prevent moisture from entering the top edge of roof flashing, such as on a chimney or wall.

course One series of shingles in a horizontal, vertical, or 45-degree angle from the eaves to the ridge. Shingles, shakes, and roll roofing are laid in successive courses.

coverage Amount of weather protection afforded by shingles or other roofing material.

cricket Small, sloped structure made of metal and designed to drain moisture away from a chimney; usually placed at the back of a chimney.

CSPE chlorosulphonated polyethylene is a single-ply roofing membrane usually made of synthetic rubber and is applied on flat or very low-slope roofs.

cupola Square or round roof structure sometimes capped by a weather vane; often used to enhance attic ventilation.

cutouts The point where the alignment notches of three-tab shingles meet to create a pattern; the watermark.

d Abbreviation for the diameter and per-penny cost of nails.

deck Planking or plywood sheathing installed over framing members; the roof surface before shingles or other roofing materials are installed.

diamond-shaped capping Cutting ridge and hip capping so that the tops of the individual caps are tapered with the lap portions hidden by the succeeding cap.

dormer Short vertical projection, usually with windows, projecting from the sloping roof and tapering into valleys on two sides.

drip edge Lightweight metal strips designed to fit snugly against rakes and eaves. Drip edge is chiefly cosmetic but the metal will provide some protection for deck edges against wind-blown moisture and some insects.

dry in Applying overlapping courses of felt to temporarily waterproof the roof deck prior to installing shingles or other roofing materials.

eaves The bottom edge of the roof deck projecting over the wall of a building.

efflorescence Chemical change causing incrustation and color variance on cement roofing products due to weathering.

EPDM Ethylene propylene diene momomer is a single-ply roofing membrane usually made of synthetic rubber and intended for application on flat or very low-slope roofs.

exposure Portion of the shingle subjected to the weather, usually 5 or $5^{1}/_{8}$ inches from the butt of one shingle to another with three-tab shingles.

face nailing Installing exposed nails heads on the roofing surface. Not a recommended practice, except as necessary for end-capping shingles.

factory cuts During the production process, some shingle manufacturers place $^{1}/_{8}$-inch cuts at the tops of shingles 6 inches from the factory edges to serve as reference marks for shingle applicators.

factory edges Shingles with untrimmed ends as produced by the manufacturer.

fascia The broad, flat outer edge of a cornice, wall, or roof projection.

felt Lightweight, asphalt-saturated building paper or inorganic fibers used to temporarily waterproof a roof deck; an underlayment for the smooth application of shingles and roll roofing over sheathing.

felt nails Square or round washers provide a 1-inch nail head to secure felt in windy conditions.

field The main body of shingles, tiles, or shakes installed after border and starter courses and below capping.

filling in Roofing a section by squaring off an angled portion to obtain long vertical runs of shingles.

flange The flat portion of a metal or plastic trim fitted over a pipe or other venting unit to waterproof the intrusion in the roof deck; sometimes called a pipe collar or vent collar.

flashing Strips of copper, aluminum, or galvanized sheet metal used along walls, dormers, valleys, and chimneys to prevent moisture seepage.

45 pattern With tree-tab shingles, application resembling steps at a 45-degree angle. Shingle cutouts align at every other course.

furring strips Lightweight wood strips applied as supplemental fastening to temporarily prevent wind damage to felt on a roof deck. Fastening base for wood roofing materials; battens.

gable Triangular end portion of a building from the eaves to the ridge of the roof.

galvanized sheet-metal roofing Metal roofing panels, approximately 3 feet wide and 10 to 13 feet long, usually installed in vertical sheets.

gauge The slot on the blade of a roofing hatchet or a notch on a pneumatic nail gun or stapler used to establish standard shingle exposure. Classification indicating the diameter of the wire used in making nails; roofing nails are designated 10 gauge.

general contractor Business or person who contracts to provide specific work for a set price and who may employ subcontractors to provide services.

Greek tile roofing Flat clay or cement tile bridged by semihexagonal caps.

high nailing Driving two or three nails near the top edge of a shingle or a series of shingles in order to properly position and align courses so that, for example, flashing can be applied. The full complement of nails is subsequently added to all high-nailed courses.

high-wind areas In the United States, regions along the Gulf Coast, the coast of Alaska, the Atlantic Coast, Guam, Hawaii, Puerto Rico, and the U.S. Virgin Islands are often subjected to 80 mph or more wind gusts.

hip pad Molded rubber or leather piece strapped or buckled over a shinglers leg and hip to provide protection against the abrasive shingle surface as the roofer sits in the standard shingler's position.

hip roof The angle formed where the sloping roof sections meet. Slopes are angled toward the center from four sides and where there are no rakes.

hook blade Used to trim or cut asphalt-fiberglass shingles, a utility knife blade equipped with a hooked cutting surface designed to grab the material as it slices.

hot-dipped galvanized nails Steel nails are placed once or twice in molten zinc hot enough to form an alloy on the outer layer of the nails.

hot-galvanized nails Zinc-coated steel nails for added weather protection.

ice dam In regions where snow and ice are common, meltwater caused by a combination of inadequate insulation and ventilation and uneven attic temperatures that refreezes near the eaves and blocks drainage into the gutter.

interlayment felt An 18-inch-wide strip of felt installed between each course of roof covering, such as wood shakes, to provide protection against the weather. The interlayment of felt is an option for wood shingles.

jack collar *See* pipe collar.

laced valley Where two internally sloping roof sections intersect, shingles are installed with the courses extending fully across both sides of the angles of the valley, with neither side trimmed at the center line of the valley, to form a channel for rain and meltwater runoff; closed valley.

laddeveyor Combination ladder and conveyor belt that lifts bundles to the roof using a gas-powered hoist, pulley, and sled.

laminated asphalt-fiberglass shingles *See* architectural shingles.

lathing Thin strips of wood, often installed with battens, used to form a latticework on which to hang tile.

limited product warranty A warranty in which roofing-product manufacturers will investigate a complaint to determine if the roofing contractor complied with all product installation procedures and requirements and then potentially offer any adjustment for premature product failure based on how long the product was in service.

local building code Construction-industry standards and practices—often adopted from the Uniform Building Code—tailored to local government agency requirements.

louver Ventilating unit with slats; usually placed at gables.

mansard roof Steeply sloped portion of a two-part roof deck, often having one or more windows, where shingles generally must be applied from a ladder or a scaffolding system.

mastic Asphalt plastic-base roof coating used for thermal insulation or waterproofing around vents and other roof obstacles; roofing cement.

match-up Where eaves border shingles are installed improperly so that the border shingle and the first-course shingle match up or align edges; a common error that could result in a series of leaks across the eaves.

mechanic's lien   A legal claim secured on property by the contractor who built, improved, or repaired the property to ensure payment for labor or materials.

metal roofing   Galvanized corrugated metal panels, stainless steel, and copper or aluminum shakes and shingles are examples of metal roof coverings.

mission tile roofing   Semicircular or barrel-shaped roofing tiles made of cement or clay.

one-tab shingle   One-third of a standard full-length asphalt-fiberglass shingle used to form a ridge or hip cap or to maintain the 45-degree step or random-spacing patterns.

open valley   Where two internally sloping roof sections intersect, shingles are trimmed at a specified width to expose metal, roll roofing, or membrane and to form a channel for rain and meltwater runoff.

organic shingles   *See* asphalt shingles.

over-the-top shingling   A roof with one layer of worn but not badly warped or curled shingles can be reroofed by applying a new layer of shingles directly over the worn layer of shingles.

peak   *See* ridge.

penny   Measure of nail length indicated by the letter d; originally an indicator of price per 100 nails.

pipe collar   A galvanized-metal, aluminum, polyurethane, lead, or copper trim fitted over the neck and base of a pipe or roof vent to waterproof the intrusion on the roof deck; sometimes called a flange or vent collar.

pitch   The incline, or slope, of a roof; the ratio of the total rise to the total width of a house measured in inches per rise per foot of run.

ply   A felt layer in a built-up roof system.

polymer roofing   Lightweight, thermoplastic-resin roofing tiles and panels resembling slates and shakes. Plastic roofing is long-lasting, walkable once installed, and requires no battens or lath as a base.

polyurethane foam roofing   Foam products from a mix of chemicals that provides a seamless, lightweight waterproof surface; used on low-slope roof applications.

purlins   Horizontal wood support strips placed between the roof frame's plate and the ridge.

rake   Edges of the roof deck running parallel to the slope.

random-spacing pattern   Pattern with cutouts aligned every six shingles using three-tab shingles.

recover *See* over-the-top shingling.

recovery fund Pool of funds provided by licensed contractors to cover homeowner losses caused by a licensed contractor who fails to adequately improve property.

ridge Horizontal junction of the two top edges of two sloping roof sections.

ridge capping *See* capping shingles.

roll roofing Asphalt-based roofing material weighing 45 to 90 pounds per roll and laid horizontally over low-pitched roofs.

roof jacks Metal supports nailed to the roof deck to provide a platform from which roofers work on steeply pitched roofs.

roofing felt *See* felt.

run Distance covered by the application of shingles in one pass of the pattern; an inclined course.

seal-down strips Factory-applied, sunlight-activated adhesive that bonds asphalt-based fiberglass shingles to the course above.

selvage-edge roll roofing Roll roofing combining 17 inches of mineral-surfaced material and 19 inches of saturated felt.

shakes Hand-split, edge-grained, textured-surface roof covering made from cedar, redwood, or cypress. Also a composite material manufactured to resemble wood. Both wood and composite shakes are designed to be thicker than wood shingles.

sheathing Plywood or planks that form the surface of the roof deck.

sheathing paper *See* felt.

shingles Roof covering made from asphalt, asphalt-fiberglass, wood, aluminum, tile, slate, or other water-shedding material. Cedar, redwood, and cypress shingles are machine cut to a smooth surface thinner than wood shakes.

short course Shingles installed horizontally with less than standard exposure but with the standard vertical alignment and application pattern. Installation of a short course usually results from the shingle pattern intersecting with an angle or a cut in the roofline.

six nailing Practice of installing six nails per three-tab shingle instead of the standard four nails. Two additional nails are positioned adjacent the inner two tabs in an attempt to better fasten shingles to roofs in geographic locations where high winds are common.

slate roofing Quarried stone, usually from Pennsylvania, Virginia, or Vermont, sometimes used as an expensive roofing option in the North and Northeast regions of the country. Colors are dictated by the chemical and mineral content of the original clay sediment.

slope Degree of incline of a roof plane usually given in inches of rise per horizontal foot of the run.

slots The open portion between tabs on asphalt-fiberglass shingles.

soffit Underside of the eaves.

Spanish tile roofing S-shaped roofing tiles made of clay or cement.

square Used to mean 100 square feet of roofing material coverage.

starter course First row of shingles or roll roofing applied at the eaves.

step flashing Usually aluminum or galvanized sheet metal cut in L-shaped pieces that are weaved at the joints between the roof surface and the roofing material; installed along walls and masonry.

straight blade Utility knife equipped with a blade that has a straight cutting surface; designed to slice material as it cuts.

straight pattern Vertical application of three-tab shingles with cutouts aligned every other shingle.

subcontractor Tradesperson hired by a general contractor or contractor for specific tasks.

tab The material between the factory cutouts on asphalt and asphalt-fiberglass shingles; a one-tab shingle.

terne An alloy of tin and lead used for plating or coating steel.

tie-in course Rejoining the courses in the correct sequence after an obstacle or intrusion, such as a chimney, interrupts the field pattern of shingles, tiles, or shakes.

tile roofing Roof covering made from clay, cement, or other water- shedding material manufactured in several weights, sizes, shapes, and colors—sometimes made to resemble shakes or shingles.

tin snips Hand shears used for cutting and trimming galvanized metal, aluminum, and other flashing and roofing materials; aviation snips.

torch roll roofing Activated by heat from a propane torch, the sealant built into the roll roofing material is designed to provide waterproof surfaces for low-slope decks and carports.

two-tab shingle Two-thirds of a standard full-length asphalt-fiberglass shingle used to maintain the 45-degree step or random-spacing patterns.

underlayment *See* felt.

Uniform Building Code Standard building-industry practices frequently adopted by local governments as requirements designed to ensure uniformity and safety in the construction of residential units. Published about every three years by the International Conference of Building Officials.

upside-down shingle   Rotating a full shingle in order to install an eaves border when a first course of shingles is started.

valley   Internal angle formed by the junction of two sloping roof sections. *See* laced valley, open valley, and Western weave valley.

vapor barrier   Material such as asphalt-saturated felt placed between the wood deck and shingles or other waterproof roof covering to retard the flow of water vapor.

vent collar   *See* flange.

waterlock   The grooved portion of cement tiles that interlock with the common edge of an adjoining tile.

watermark   Line where double coverage of roofing material laps courses. *See* cutout.

Western weave valley   Asphalt and asphalt-fiberglass shingles are installed with the courses extending fully across one side of the angle of a valley and the opposite courses from the other angled roof section. The shingles are overlapped, then trimmed at the center line of the valley.

whole-house fan   Ventilator usually installed in an attic floor to exhaust air from the entire house.

wing   Roof section broadly extended or projecting at an angle from the main building.

wood shake   *See* shake.

wood shingle   *See* shingle.

Index